META-ANALYSIS of BINARY DATA USING PROFILE LIKELIHOOD

CHAPMAN & HALL/CRC
Interdisciplinary Statistics Series

Series editors: N. Keiding, B.J.T. Morgan, C.K. Wikle, P. van der Heijden

Published titles

Published titles

Interdisciplinary Statistics

META-ANALYSIS of BINARY DATA USING PROFILE LIKELIHOOD

Dankmar Böhning
University of Reading
UK

Ronny Kuhnert
Robert Koch-Institut
Berlin, Germany

Sasivimol Rattanasiri
Ramathibodi Hospital
Bangkok, Thailand

CRC Press
Taylor & Francis Group
Boca Raton London New York

CRC Press is an imprint of the
Taylor & Francis Group, an **informa** business
A CHAPMAN & HALL BOOK

CRC Press
Taylor & Francis Group
6000 Broken Sound Parkway NW, Suite 300
Boca Raton, FL 33487-2742

First issued in paperback 2019

© 2008 by Taylor & Francis Group, LLC
CRC Press is an imprint of Taylor & Francis Group, an Informa business

No claim to original U.S. Government works

ISBN-13: 978-1-58488-630-3 (hbk)
ISBN-13: 978-0-367-38757-0 (pbk)

Library of Congress Cataloging-in-Publication Data

Böhning, Dankmar.
   Meta-analysis of binary data using profile likelihood / Dankmar Böhning,
Ronny Kuhnert, and Sasivimol Rattanasiri.
      p. cm. -- (Interdisciplinary statistics series)
   Includes bibliographical references and index.
   ISBN 978-1-58488-630-3 (alk. paper)
   1. Meta-analysis. 2. Medicine--Research--Evaluation. 3. Social
sciences--Statistical methods. I. Kuhnert, Ronny. II. Rattanasiri, Sasivimol. III.
Title.

R853.M48B64 2008
610.72'7--dc22   2007049869

Visit the Taylor & Francis Web site at
http://www.taylorandfrancis.com

and the CRC Press Web site at
http://www.crcpress.com

Contents

Preface

With the increasing number of empirical studies on a question of interest, the topic of systematic reviews and meta-analysis is becoming of sustainable interest. Ideally, all studies involved in a meta-analysis should have all individual patient data available. This situation is sometimes also called *meta-analysis of pooled data*. On the other extreme only an effect measure such as the odds ratio is available from published studies. Clearly, in the latter situation fewer options are available. For example, the effect measure cannot be changed in the meta-analysis. This book considers a compromise situation that is frequently available: from the published literature a 2×2 table is available containing the successes and failures in the treatment arm as well as the successes and failures in the control arm. Although also in this situation individual patient data are *not* available, there is considerably more information present than in a meta-analysis based entirely on reported effect measures. We call this situation a *meta-analysis with individually pooled data* (MAIPD). In MAIPD different effect measures can be computed such as the odds ratio, risk ratio, or risk difference accompanied by a standard error, which is more reliably computed than the one conventionally available in a meta-analysis from published effect measures. The book is devoted to the analysis and modeling of a MAIPD.

The book is outlined as follows. In chapter 1, the meta-analytic situation of a MAIPD is introduced and illustrated with several examples. In chapter 2 the basic model is introduced including the profile likelihood method and a discussion of it under homogeneity. In section 3, the model is extended to cope with unobserved heterogeneity, which is captured by means of a nonparametric mixture leading to the nonparametric mixture profile likelihood. The gradient function is introduced and the nonparametric profile maximum likelihood estimator (PNMLE) is characterized. The latter can be computed by means of the EM algorithm with gradient function update (EMGFU). This ends section 3. Section 4 provides modeling of covariate information. Elements of log-linear modeling are used and ways for finding the profile maximum likelihood estimator, including standard errors, are provided. In chapter 5 alternative approaches to the profile likelihood method are discussed including an approximated likelihood model on the basis of the normal distribution as well as the multilevel approach. Chapter 6 discusses ways to model covariate information and unobserved heterogeneity simultaneously. Chapter 7 gives an illustration of the program CAMAP, which has been developed for the analysis of a MAIPD and is accompanying and supplementing this book as freely downloadable software. Chapter 9 approaches the problem of quantifying heterogeneity in a MAIPD. Not only it is important to decide whether there is or there is not heterogeneity, but also, if there is, how large is the amount of heterogeneity in the MAIPD. Chapter 10 shows that the methodology can also be applied to a surveillance problem, here the surveillance of scrapie in Europe, and is not restricted to clinical trials.

The idea for this book started while Dankmar Böhning was still professor at the Charité Medical School Berlin (Germany) where he had been exposed over many years to the problems of meta-analysis and systematic reviews. The various chapters that are contained in this book grew over time in which the various ways of modeling were explored. It should be mentioned that various projects were supported and generously funded by the German Research Foundation (DFG), which accompanied the principal investigator over many years with interest and support. The coauthors of this book, Sasivimol Rattanasiri and Ronny Kuhnert, developed their dissertational projects out of the theme of this book, in fact, their dissertations became part of it. Clearly, not all aspects of meta-analysis could be covered. The problem of *publication bias* is not discussed. We felt that publication bias is more relevant for meta-analysis based upon published literature, whereas clinical trials with planned and registered studies included in the meta-analysis are less prone to this form of bias. Also, Bayesian methods are used only in an empirical Bayesian sense, for example when using the maximum posterior allocation rule for allocation of the individual studies into their associated clusters. We felt that a full Bayesian approach would have increased complexity, size, and timeframe of the book (as well as the associated software) considerably, which we preferred to avoid.

The book should be of interest for almost everyone interested in meta-analysis. Many chapters contain new developments not available in the current literature. This is particularly true for chapters 4, 6, 8, and 10. The book might be well used as a supplementary textbook in any graduate course on meta-analysis with emphasis on applied statistics.

Finally, we, Dankmar Böhning, Ronny Kuhnert, Sasivimol Rattanasiri, would like to thank deeply:

- the German Research Foundation (DFG) for its enormous financial and academic support,
- professor em. Dr. Frank-Peter Schelp for generously supporting this project while we were working under his leadership in the Institute for International Health of the Charité Medical School Berlin,
- Victor Del Rio Vilas (Veterinary Laboratory Agency, UK) for pointing out the potential in using the methodology of a MAIPD in surveillance problems,

- professor Heinz Holling (Münster, Germany) for his long-lasting interest and cooperation in many areas of applied statistics, but in particular in meta-analysis,
- professors Anne and John Whitehead (Lancaster, UK) for the many discussions we had on meta-analysis and other questions of interest,
- Dr Mike Dennett (Reading, UK) for taking the burden of head of section,
- the publisher Chapman & Hall/CRC, in particular, Rob Calver and his team for his challenged patience with the delivery of this book,
- our children, Anna-Siglinde and Laura, for the play time we took away from them,
- our spouses Tan, Katja, and Prasan for their tolerance they showed with this seemingly endless project,
- and our parents for their love.

Dankmar Böhning (Reading, UK)

Ronny Kuhnert (Berlin, Germany)

Sasivimol Rattanasiri (Bangkok, Thailand)

February 2008

Abbreviations

AIC	—	Akaike Information Criterion
AL	—	Approximate Likelihood
AS	—	Abattoir Stock
BCG	—	Bacillus Calmette-Guérin
BIC	—	Bayesian Information Criterion
BHAT	—	Beta-Blocker Heart Attack Trial
BSE	—	Bovine Spongiform Encephalopathy
CALGB	—	Cancer and Leukemia Group B
CAMAP	—	Computer-Assisted Meta-Analysis with the Profile Likelihood
CI	—	Confidence Interval
EM	—	Expected Maximization
EMGFU	—	EM Algorithm with Gradient Function Update
EU	—	European Union
FS	—	Fallen Stock
IIID		Ischaemic Heart Disease
log	—	Natural Logarithm
MAIPD	—	Meta-Analysis for Individually Pooled Data
MAP	—	Maximum Posterior Allocation Rule
MC	—	Multicenter
MH	—	Mantel-Haenszel
MHE	—	Mantel-Haenszel Estimate
ML	—	Maximum Likelihood
MLE	—	Maximum Likelihood Estimator
ML	—	Multilevel
NRT	—	Nicotine Replacement Therapy
PL	—	Profile Likelihood
PLRT	—	Profile Likelihood Ratio Test
PML	—	Profile Maximum Likelihood
PMLE	—	Profile Maximum Likelihood Estimator
PNMLE	—	Profile Nonparametric Maximum Likelihood Estimator
OR	—	Odds Ratio
RP	—	Representativeness
RR	—	Relative Risk
TB	—	Tuberculosis
TSE	—	Transmissible Spongiform Encephalopathies

CHAPTER 1

Introduction

1.1 The occurrence of meta-analytic studies with binary outcome

The present contribution aims to provide a *unifying* approach to modeling treatment effect in a meta-analysis of clinical trials with binary outcome. In recent years, meta-analysis has become an essential method used to provide more reliable information on an intervention effect. Additionally, it has been demonstrated to provide a powerful statistical tool to analyze and potentially combine the results from individual studies. Numerous international publications have demonstrated the quality and the common practicability of meta-analysis (see for example Cooper and Hedges (1994), Sutton et al. (2000), DuMouchel and Normand (2000), Jones (1992), or Greenland (1994)). Important for our situation here is the availability of the number of events x_i^T (x_i^C) and the person-time under risk n_i^T (n_i^C) (total of time every person spent under risk) in the treatment arm (control arm) for each clinical trial i involved in the meta-analysis of a total of k studies. If all persons spend identical time under risk n_i^T is equivalent to the sample size, and the same is true for the control arm. We call this situation of meta-analysis a *meta-analysis using individually pooled data* (MAIPD). Table 1.1 shows the principal layout of the required information.

Table 1.1 *Principal layout of the required information in a MAIPD for study i*

Arm	Number of Deaths	Person-Time
Treatment	x_i^T	n_i^T
Control	x_i^C	n_i^C

Frequently, data in the format of Table 1.1 are presented on the basis of trial sizes (so that each person contributes the identical person-time). However, we believe that the concept of person-time is far more general and we prefer to lay out the concept for situations that cover varying person-times.

For illustration, consider the data given in Table 1.2 taken from Petitti (1994). The table contains outcome data from a large randomized clinical trial of antiplatelet treatment for patients with a transient ischemic attack or ischemic

1

stroke identified by the Antiplatelet Trialists' Collaboration (1988) as eligible for their meta-analysis. An event was defined as first myocardial infarction, stroke, or vascular failure. In this case, all patients had a similar study period as their time being under risk so that the person-time n of each trial arm corresponds to the number of patients under risk, here 1,250. The data shown in Table 1.2 came from the European Stroke Prevention Study Group (1987).

Table 1.2 *Data from a randomized trial of antiplatelet therapy for treatment of transient ischemic attack of stroke*

Arm	Number of Deaths	Number at Risk
Treatment	182	1,250
Control	264	1,250

A quick analysis with the package STATA (StataCorp. (2005)) shows that the estimate of the risk ratio, defined as the ratio of x^T/n^T to x^C/n^C, is given as $\frac{182/1,250}{264/1,250} = 0.6894$ with a 95% confidence interval of $0.5806 - 0.8186$, indicating a significant preventive effect of the treatment. If Table 1.2 contains all the evidence available for assessing the question of the effect of treatment, then this is all that could be done on this level of information. However, Petitti (1994) provides data from a second randomized trial containing the outcome date from the United Kingdom Transient Ischemic Attack Aspirin Trial (unpublished), which we provide here as Table 1.3.

Table 1.3 *Data from a second randomized trial of antiplatelet therapy for treatment of transient ischemic attack of stroke*

Arm	Number of Deaths	Number at Risk
Treatment	348	1,621
Control	204	814

An analysis of the data of Table 1.3 similar to the analysis of Table 1.2 delivers an estimate of the risk ratio $\frac{348/1,621}{204/1814} = 0.8566$ with a 95% confidence interval of $0.7366 - 0.9962$, indicating a significant, but more borderline preventive effect of the treatment. This change of effect size might have substantial reasons or might be purely due to chance. It definitely required further and deeper analysis.

Hence, having simply another trial available raises already several questions:

- a) How can the trial-specific risk ratio estimates be combined?
- b) Is the combination an efficient use of information?

- c) Is the combination itself valid?

Table 1.4 shows the results of an analysis of these data using the package STATA (StataCorp. (2005)). A Mantel-Haenszel estimate is calculated as

$$\hat{\theta}_{MH} = \sum_i w_i \hat{\theta}_i / \sum_i w_i,$$

where $\hat{\theta}_i$ are the study-specifc risk ratio estimates of studies $i = 1, 2$ and the Mantel-Haenszel weights are given as $w_i = n_i^T x_i^C / (n_i^T + n_i^C)$. The Mantel-Haenszel approach is a well-known approach (see Woodward (1999)) and, as it will be established later, a reasonably efficient approach. To provide a better intuitive understanding we write the Mantel-Haenszel estimator as

$$\hat{\theta}_{MH} = \frac{\sum_i x_i^T n_i^C / (n_i^T + n_i^C)}{\sum_i x_i^C n_i^T / (n_i^T + n_i^C)},$$

which follows the Mantel-Haenszel construction rule of *taking sums before ratios*. The estimator is remarkably robust against the occurrences of zero events in the individual studies. As Table 1.4 shows, adjusting for a study effect by means of stratifying over studies will prevent a confounding effect which could occur when simply considering the pooled estimate of both studies. Comparing the crude (0.8142) and study-adjusted effects (0.7742) we observe a very small confounding effect of study showing the simply pooled estimate of relative risk would underestimate the true underlying relative risk.

Table 1.4 **STATA**-*output (StataCorp. (2005)) from the two randomized trials of antiplatelet therapy for treatment of transient ischemic attack of stroke*

Study	RR	95% Conf. Interval	M-H Weight
Table 1.2	0.6894	0.5806 - 0.8186	132.00
Table 1.3	0.8566	0.7366 - 0.9962	135.80
Crude	0.8142	0.7288 - 0.9095	
M-H combined	0.7742	0.6912 - 0.8672	
Test of homogeneity (M-H) $\chi^2(1)$ =3.478 Pr> $\chi^2(1) = 0.0622$			

The final issue raised above was if the study-specific relative risk estimates can be validly combined via the Mantel-Haenszel method. This is approached by the test of homogeneity which compares the study-specific estimates of relative risk and is found in Table 1.4 to be 3.478 which is not significant on a chi-square distribution with 1 degree of freedom. We conclude that there is no evidence for rejecting a relative risk common to the two studies so that the Mantel-Haenszel summary measure appears reasonable.

A more modeling-type approach would consider a *log-linear model* for the observed number of events X_s^t such as

$$E(X_s^t) = \exp\{n_s^t + \alpha + \beta t + \gamma s\} \tag{1.1}$$

where β is the treatment effect and t is a binary variate indicating treatment $(t = 1)$ or control $(t = 0)$ and s is a binary variate indicating study two $(s = 1)$ or study one $(s = 0)$. One of the specific characteristics is that the person-time n_s^t occurs as an *offset* term in (1.1). Estimation by means of maximum likelihood could be handled by considering (1.1) as a *generalized linear model* where a Poisson is taken as an error distribution for the number of events X_s^t. Many modern packages such as STATA (StataCorp. (2005)) and SAS among many others offer these procedures nowadays in a user-friendly way. The model (1.1) could be extended to include a further interaction term $(\beta\gamma)$

$$E(X_s^t) = \exp\{n_s^t + \alpha + \beta t + \gamma s + (\beta\gamma)st\} \tag{1.2}$$

so that the hypothesis of effect homogeneity could be more elegantly approached by considering a likelihood-ratio test in forming a ratio of likelihoods of models (1.2) and (1.1), respectively. If the test is significant, then this simply means that the treatment effect changes when the study changes. As a result, the treatment effect will depend on which study is considered. Therefore, where is the problem the reader may ask?

The situation becomes more complicated if there are more and more studies considered. Technically, it is easily possible to introduce for every new study a further indicator variable. However, this masks a more structural problem, namely the fact that each new study also requires at least one additional parameter in the model. If the number of studies is considered as sample size, then the number of parameters in the model increases linearly with the sample size. This introduces a *Neyman-Scott* problem which raises the question of parameter estimation in the presence of infinitely many nuisance parameters (see Neyman and Scott (1948)). For our situation the model (1.3) could be generalized to

$$E(X_j^t) = \exp\{n_j^t + \alpha + \beta t + \gamma_j s_j\} \tag{1.3}$$

where s_j are now binary variables indicating that the data are from study j, $j = 1, \cdots, k$. To avoid overparameterization we set $\gamma_1 = 0$. Note that the k parameters $\alpha, \gamma_2, \cdots, \gamma_k$ correspond to the baseline risks in k studies.

Kiefer and Wolfowitz (1956) provide a solution to the Neyman-Scott problem by considering the sequence of nuisance parameters to arise from a distribution in which case a *consistent* estimation of this distribution becomes possible. In a simple parametric approach, this leads to assume a normal random effect for the study effects, $\gamma_2, \cdots, \gamma_k \sim N(0, \sigma_s^2)$. Other parametric distributions such as the Gamma-distribution are also possible as random effects distribution (see Biggerstaff and Tweedie (1997) or Berry (1998)). If the outcome variable is a normal response we are in the area of *mixed models*, see for example Brown and Prescott (1999); Demidenko (2004). If the random effects distribution is left *unspecified*, a *nonparametric maximum likelihood estimator* can be constructed, an approach followed by several authors including Aitkin (1999a); Aitkin (1999b); Böhning (2000); Lindsay (1995).

In this setting of several, potentially numerous studies of treatment effect on binary outcome a different approach will be followed. Instead of estimating the distribution of nuisance parameters (risk baseline parameters), the nuisance parameters are eliminated before further modeling on the treatment effect parameter is considered (see also Aitkin (1998)). The benefit of the approach - as will be worked out in Chapter 5 - lies in focusing the inference on the parameter of interest, hence, providing a more powerful approach in comparison to other methods that simultaneously model *nuisance* and *treatment effect parameters*. Murphy and Van der Vaart (2000) provide arguments that profile likelihoods can be treated just as likelihoods, despite earlier critics on the profile likelihood method who focused on the fact that the variability due to replacing the nuisance parameters by estimates is not adjusted for in the construction of the profile likelihood. We will demonstrate these arguments and show in Chapter 5 that the profile likelihood method works best, at least for the setting of this monograph, among all other methods considered, and currently we see more the difficulties of the profile likelihood in its application in models where the profile likelihood is technically more elaborate in its construction.

Suppose we now have the number of events x_i^T and x_i^C for treatment and control arm, respectively, in study i available, $i = 1, \cdots, k$ with associated person-times n_i^T and n_i^C. As an example we consider the MAIPD of the effect of beta-blockers for reducing mortality after myocardial infarction (Yusuf et al. (1985)). The individual-study pooled patient data for 22 studies are provided in Table 1.5. The effects (expressed as log-relative risks) with their associated 95% confidence intervals are given in Figure 1.1. Several aspects can be explored from this graph.

- It can easily be seen which of the studies have positive and which have negative effects.

- Significant positive effects can be easily identified by finding those studies with confidence intervals above the no-effect line, the horizontal line at zero. Here, there are none.

- Significant negative effects can also be easily identified by finding those studies with confidence intervals below the no-effect line. These are the four studies: 7, 10, 21, and 22.

- Some studies show strong negative effects (stronger than the studies 7 and 10 which had a significant negative efffect) but are *not* significant, since there is a large variance associated with these studies due to a small size. This refers to studies 2, 3, 6, and 13.

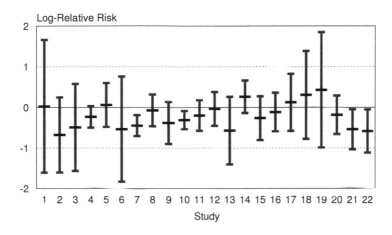

Figure 1.1 *Effect of beta-blocker for reducing mortality after myocardial infarction (Yusuf et al. (1985)) in 22 studies expressed as log-relative risk with 95% CI*

These aspects become visible when constructing graphs such as Figure 1.1 which are important explorative tools. These need to be analyzed deeper by appropriate statistical methods.

Note that frequently a MAIPD could only be done on the basis of the published literature (for a collection of numerous meta-analyses of this type see Kuhnert (2005)). However, individual patient data are typically *not* available from a MAIPD and are difficult, not to say usually impossible to retrieve. This usually does not allow modeling on the individual patient or person level. Note further that in contrast to conventional meta-analysis where typically an effect measure accompanied by a variance estimate is available (see, for example van Houwelingen et al. (2002)), a MAIPD offers more choices for the analyst, for example in choosing different effect measures such as relative risk, risk difference, and number-needed-to-treat with accompanying variances. In this contribution we will exploit another option of MAIPD, namely we will use the profile likelihood method to eliminate the nuisance baseline parameter and consecutively will base inference on the profile likelihood. In particular, the method is able to cope with the two forms of heterogeneity that can occur in a MAIPD: the *baseline heterogeneity* that arises in the control arm and the *effect heterogeneity* that arises from potentially different treatment effects in the various centers.

1.2 Meta-analytic and multicenter studies

Another issue concerns the connection of MAIPD to the analysis of multicenter studies, frequently mentioned in the literature. Note that in a multicenter clinical trial the methodology used in MAIPD is validly applicable as well. However, in the multicenter setting individual patient data are usually available and will offer more analysis options. On the other hand, multicenter study data are often neither published, nor are they made available to the interested reader for numerous reasons, though center-specific information might be found in the corresponding publication. Therefore, the methodology of MAIPD might be very useful in this situation as well. Consequently, MAIPD is a common situation in multicenter clinical trials (see also Aitkin (1999b)) and all models discussed here are also applicable in multicenter studies. Bearing this in mind, one should see the clear differences between MAIPD in general on the one hand and a multicenter trial on the other hand. For example, one difference between the two is the usually stronger protocol restrictions in the case of multicenter studies. Therefore, the baseline heterogeneity in a MAIPD can be expected to be larger.

In this contribution the focus is on binary outcome of the trial such as survival (yes/no), improvement of health status (yes/no), occurrence of side effects (yes/no) to mention a few potential binary outcomes. Even if the outcome measure is continuous (such as blood pressure) it is often categorized into two possible values. Frequently, analysis in a clinical trial is done on a time-to-event basis. Although this might be a desirable objective, most MAIPD we looked at presented a binary outcome measure. Therefore, for the time being, we will concentrate here on modeling binary outcomes.

Other examples include diagnostic procedures that frequently result in continuous measures. However, the outcome is almost uniquely represented in terms of positive result or negative result.

Multicenter studies occur in numerous ways and are executed quite frequently. It is said that "there are currently thousands of active multicenter trials designed to evaluate treatment or prevention strategies" (Bryant et al. (1998)). Among the reasons for mounting a multicenter study are the *need to recruit patients at a faster rate*, the *need to find patients with a rare disease or condition*, or the *desire to increase the generalizability of effect*, because multicenter studies will more likely include heterogenous populations. Pocock (1997) points out that the collaboration of clinical scientists in a multicenter study should lead to increased standards in the design, conduct, and interpretation of the trial. Often the center represents a clinical, medical, or public health institution in which the clinical trial takes place.

Furthermore, it is assumed that each trial is competitive in that it compares two (or more) trial arms, here denoted as *treatment* and *control* arm. A typical setting is provided in Table 1.5 in which the data of a MAIPD of clinical trials are presented to study the effect of beta-blockers for reducing mortality after myocardial infarction (Yusuf et al. (1985)). A manifold collection of MAIPDs in various application fields is provided in the Cochrane Library (2005).

Table 1.5 *Data illustration of a MAIPD for studying the effect of beta-blocker for reducing mortality after myocardial infarction (Yusuf et al. (1985)), data contain number of deaths x_i and person-time n_i for treatment (T) and control arm (C) as well as risk ratio estimate (last column)*

Trial	Treatment		Control		Relative Risk
i	x_i^T	n_i^T	x_i^C	n_i^C	$\hat{\theta}_i$ (95%CI)
1	3	38	3	39	1.0263 (0.2207–4.7716)
2	7	114	14	116	0.5088 (0.2132–1.2140)
3	5	69	11	93	0.6127 (0.2231–1.6825)
4	102	1533	127	1520	0.7963 (0.6197–1.0233)
5	28	355	27	365	1.0663 (0.6415–1.7722)
6	4	59	6	52	0.5876 (0.1754–1.9684)
7	98	945	152	939	0.6406 (0.5053–0.8122)
8	60	632	48	471	0.9316 (0.6497–1.3357)
9	25	278	37	282	0.6854 (0.4243–1.1072)
10	138	1916	188	1921	0.7360 (0.5963–0.9083)
11	64	873	52	583	0.8219 (0.5789–1.1671)
12	45	263	47	266	0.9684 (0.6678–1.4041)
13	9	291	16	293	0.5664 (0.2544–1.2610)
14	57	858	45	883	1.3036 (0.8920–1.9050)
15	25	154	31	147	0.7698 (0.4783–1.2390)
16	33	207	38	213	0.8936 (0.5840–1.3673)
17	28	251	12	122	1.1341 (0.5976–2.1524)
18	8	151	6	154	1.3598 (0.4833–3.8259)
19	6	174	3	134	1.5402 (0.3924–6.0462)
20	32	209	40	218	0.8345 (0.5459–1.2756)
21	27	391	43	364	0.5846 (0.3692–0.9256)
22	22	680	39	674	0.5591 (0.3352–0.9326)

Interest lies in measuring the effect of treatment, frequently accomplished by means of the risk ratio $\theta = p^T/p^C$ where p^T and p^C are the risks of an event under treatment and control, respectively. Nowadays, it is widely accepted that a simple, overall estimate of the crude risk ratio estimate

$$\hat{\theta}_{crude} = \frac{\left(\sum_{i=1}^{k} x_i^T\right) / \left(\sum_{i=1}^{k} n_i^T\right)}{\left(\sum_{i=1}^{k} x_i^C\right) / \left(\sum_{i=1}^{k} n_i^C\right)}$$

is by no means a sufficient description of the available data - unless effect homogeneity is established. Mainly, three reasons are responsible for this perspective.

- a) The simple estimate ignores a potential center effect. In Table 1.5 most trials show a beneficial effect of treatment with an overall beneficial estimate of treatment effect as $\hat{\theta}_{crude} = 0.79$, though treatment effect is not always protective as can be seen in trials 14, 18, or 19. Note also that most of the trials experience a nonsignificant effect.

- b) To avoid confounding by trial (the center effect) a stratified analysis is usually recommended. Then, the question arises, which sources are responsible for the deviations of trial-specific estimates of treatment effect from the adjusted overall estimate. Are these pure sources of random error or do other sources such as unobserved covariates, here summarized under *unobserved heterogeneity*, also contribute to the observed error. A controversial example and discussion on effect heterogeneity is given in Horwitz et al. (1996) in which 21 centers show beneficial and 10 centers harmful effects (see also Table 1.12).

- c) Finally, the simple estimate ignores potentially observed covariate information which should be taken into account. Maybe age and gender distributions varied from trial to trial. Maybe randomization failed in some trials. Maybe the background population was different from trial to trial.

This contribution aims to achieve a solid modeling of the three before mentioned situations accompanied by easy-to-use inferences and algorithms which will allow the clinician to analyze MAIPDs in an up-to-date method.

1.3 Center or study effect

It might be tempting to ignore the fact that study data are available for different centers. Indeed, it is possible to collapse data over all centers to achieve a simple two-by-two table from which the effect estimate could be computed simply as $\hat{\theta}_{crude} = \frac{\left(\sum_{i=1}^{k} x_i^T\right)/\left(\sum_{i=1}^{k} n_i^T\right)}{\left(\sum_{i=1}^{k} x_i^C\right)/\left(\sum_{i=1}^{k} n_i^C\right)}$.

Though this is tempting, past experiences have shown that calculating a *crude* risk ratio as above may lead to quite biased estimates. In fact, various confounding situations can arise: the true effect might be overestimated (inflation) or underestimated (masking), or the center might work as an effect modifier. Hence, it is advisable to take the center effect as a potential confounder into account. In Table 1.6 the crude risk ratio is 1.74, well in the range of the center-specific risk ratios, and the center does not appear to be a confounder for this multicenter study. In another application, Arends et al. (2000) investigated the treatment of cholesterol lowering levels on mortality from coronary heart disease (see Table 1.6). Here, the crude risk ratio is 1.0770, whereas the Mantel-Haenszel adjusted risk ratio, $\hat{\theta}_{MH} = \frac{\sum_{i=1}^{k} x_i^T n_i^C / n_i}{\sum_{i=1}^{k} x_i^C n_i^T / n_i}$ with $n_i = n_i^T + n_i^C$, is 0.9708. This moves an elevated risk ratio to the preventive side, as can be expected from the nature of the treatment. This example underlines the importance of considering the center effect in all analyses.

In addition, another aspect might be worth mentioning. Whereas none of the center-specific risk ratio estimates in the Lidocaine trial (see Table 1.7) confirms significantly the damaging effect of prophylactic use of Lidocaine, a center-adjusting estimator like the Mantel-Haenszel estimator will provide a significant effect $\hat{\theta}_{MH} = 1.73$ with a 95% CI (1.03, 2.92). Hence, it is desirable to seek optimal and valid ways to combine available information.

1.4 Sparsity

Another aspect concerns sparsity. Frequently, in a MAIPD the observed data experience *sparsity*. The data are called *sparse* if the observed event counts are close to zero, occasionally in fact identical to zero. This can occur because the event risks are very small, so that even with a large trial sparsity has to be expected. Or, the center sizes are so small (potentially because patient recruitment is extremely difficult) that even with large event risks the occurrences of low frequency counts including zero counts are likely. An example of this nature is provided in Table 1.8 where data on a MAIPD investigating treatment in cancer patients (Cancer and Leukemia Group, Cooper et al. (1993)) are provided. The data in Table 1.8 are from the Cancer and Leukemia Group B (CALGB) randomized clinical trial comparing two chemotherapy treatments with respect to survival (lived/died by the end of the study) in patients with multiple myeloma (Cooper et al. (1993)). A total of 156 eligible patients was accrued in the 21 centers. In contrast to Table 1.5, the data in Table 1.8 are experiencing *sparsity*, meaning that the number of events and/or the number under risk is small, leading to potentially many zero events in the centers as happens in centers 5, 10, 12, and 20.

In a sparse MAIPD, the investigation of center-effect heterogeneity is particularly difficult, since center-specific risk ratio estimators can only be estimated with large uncertainty. In addition, the construction of a risk ratio estimator under homogeneity needs to be done with careful consideration. Here, the profile method turns out to be beneficial.

Table 1.6 *Outcome data of a meta-analysis of Davey Smith et al. (1993) on the effect of cholesterol lowering treatment on mortality from coronary heart disease (following Arends et al. (2000))*

Trial	Treatment		Control	
i	x_i^T	n_i^T	x_i^C	n_i^C
1	28	380	51	350
2	70	1250	38	640
3	37	690	40	500
4	2	90	3	30
5	0	30	3	30
6	61	1240	82	1180
7	41	1930	55	890
8	20	340	24	350
9	111	1930	113	1920
10	81	1240	27	410
11	31	1140	51	1140
12	17	210	12	220
13	23	210	20	230
14	0	90	4	170
15	1450	38620	723	19420
16	174	1350	178	1330
17	28	890	31	860
18	42	1970	48	2060
19	4	150	5	150
20	37	2150	48	2100
21	39	1010	28	1120
22	8	100	1	50
23	5	340	7	340
24	269	4410	248	4390
25	49	3850	62	3740
26	0	190	1	190
27	19	1510	12	1560
28	68	13850	71	13800
29	46	10140	43	10040
30	33	5910	3	1500
31	236	27630	181	27590
32	0	100	1	100
33	1	20	2	30

Table 1.7 *Outcome data for prophylactic use of Lidocaine after heart attack (AMI)*
(Hine et al. (1989), following Normand (1999))

Trial	Treatment		Control		Relative Risk
i	x_i^T	n_i^T	x_i^C	n_i^C	$\hat{\theta}_i(95\%\text{CI})$
1	2	39	1	43	2.21 (0.21–23.4)
2	4	44	4	44	1.00 (0.27–3.75)
3	6	107	4	110	1.54 (0.45–5.31)
4	7	103	5	100	1.36 (0.45–4.14)
5	7	110	3	106	2.25 (0.60–8.47)
6	11	154	4	146	2.61 (0.85–8.01)

1.5 Some examples of MAIPDs

In this section, we present a few more examples to illustrate the manifold
applications of meta-analytic studies with binary outcome.

1.5.1 Selective decontamination of the digestive tract and risk of respiratory tract infection

Turner et al. (2000) use a MAIPD on the effect of selective decontamination
of the digestive tract on the risk of respiratory tract infection (see Table 1.9).
Patients in intensive care units were randomized to receive treatment by a
combination of nonabsorbable antibiotics or to receive no treatment (Selective Decontamination of the Digestive Tract Trialists' Collaborative Group
(1993)).

1.5.2 Hypertension and cardiovascular mortality

Hoes et al. (1995) published a meta-analysis of clinical trials (see Table 1.10)
in which a drug treatment was compared to either a placebo or to no treatment with respect to (cardiovascular) mortality in middle-aged patients with
mild to moderate hypertension. In this MAIPD, varying person-times occur.
Therefore, the n_i^T and n_i^C no longer correspond to the trial size.

1.5.3 Multifarm study on Neospora infection

The data in Table 1.11 stem from a multifarm study on cows with and without
Neospora infection. A case is defined to be a cow with a history of calf-abort.
The importance lies in the fact the infection is rather widespread and the

Table 1.8 *Outcome data for treatment group of a multicenter clinical trial (with high sparsity) (Cancer and Leukemia Group, Cooper et al. (1993))*

Trial	Treatment		Control	
i	x_i^T	n_i^T	x_i^C	n_i^C
1	1	3	3	4
2	8	11	3	4
3	2	3	2	2
4	2	2	2	2
5	0	3	2	2
6	2	3	1	3
7	2	3	2	2
8	4	4	1	5
9	2	3	2	2
10	2	3	0	2
11	3	3	3	3
12	0	2	2	2
13	1	5	1	4
14	2	4	2	3
15	4	6	2	4
16	3	9	4	12
17	2	3	1	2
18	1	4	3	3
19	2	3	1	4
20	0	2	0	3
21	1	5	2	4

establishment of an effect of the infection onto the outcome could have quite an impact for the health of herds on farms.

1.5.4 Beta-Blocker Heart Attack Trial (BHAT)

The authors Horwitz et al. (1996) were interested to illustrate some of the considerable heterogeneity among the 31 centers in the BHAT study (see Table 1.12). Distinct from the comparison of mortality rates between groups of centers is the wide range in mortality by treatment group across the centers. Among 21 centers (the 1st to the 21st center) whose results favored propranolol, mortality rates in patients randomized to propranolol ranged from 0 to 13%, and from 6 to 21% in the patients randomized to receive a placebo. A similar wide variation in mortality rates was noted for patients randomized to

Table 1.9 *Respiratory tract infections in treatment and control group of 22 trials following Turner et al. (2000)*

Trial	Treatment		Control	
i	x_i^T	n_i^T	x_i^C	n_i^C
1	7	47	25	54
2	4	38	24	41
3	20	96	37	95
4	1	14	11	17
5	10	48	26	49
6	2	101	13	84
7	12	161	38	170
8	1	28	29	60
9	1	19	9	20
10	22	49	44	47
11	25	162	30	160
12	31	200	40	185
13	9	39	10	41
14	22	193	40	185
15	0	45	4	46
16	31	131	60	140
17	4	75	12	75
18	31	220	42	225
19	7	55	26	57
20	3	91	17	92
21	14	25	23	23
22	3	65	6	68

placebo and propranolol in the last 10 centers (the 22nd to the 31st center) favoring placebo.

1.6 Choice of effect measure

Frequently, medical statisticians engage in the question of which measure should be chosen to evaluate the effect of treatment (or exposure). Although there seems no ultimate data-driven solution for this question, potential evidence for a measure can be found in the following way. Suppose that one agrees with the paradigm that a model is "better" if it has higher agreement with the observed data than another model. In fact, one wants to avoid the artifact *to search for an explanation of heterogeneity in the effect if this effect is due to the choice of a wrong effect measure*. To illustrate consider the situation of Figure 1.2 where a case is made for constant risk ratio, but increasing risk difference.

Table 1.10 *Number of deaths and total number of person-years for treatment and control of 12 randomized trials in mild to moderate hypertension in the meta-analysis of Hoes et al. (1995)*

Trial	Treatment		Control	
i	x_i^T	n_i^T	x_i^C	n_i^C
1	10	595.2	21	640.2
2	2	762.0	0	756.0
3	54	5635.0	70	5600.0
4	47	5135.0	63	4960.0
5	53	3760.0	62	4210.0
6	10	2233.0	9	2084.5
7	25	7056.1	35	6824.0
8	47	8099.0	31	8267.0
9	43	5810.0	39	5922.0
10	25	5397.0	45	5173.0
11	157	22162.7	182	22172.5
12	92	20885.0	72	20645.0

Table 1.11 *Data from multifarm study on abort history of cows with and without Neospora infection, "T" serologic positive, "C" negative (Greiner (2000))*

Trial	Treatment		Control	
i	x_i^T	n_i^T	x_i^C	n_i^C
1	6	19	0	14
2	4	9	0	6
3	2	2	1	6
4	6	15	3	34
5	4	8	1	33
6	3	7	1	18
7	3	4	1	27

Now, the simplest model in the setting of Chapter 1 is the *homogeneity* model that is for the relative risk

$$p_i^T / p_i^C = \theta \text{ for all } i \text{ or } p_i^T = p_i^C \theta,$$

a straight line in the $(p_i^T, p_i^C)'$-plane through the origin with slope θ. For the risk difference, homogeneity implies that $p_i^T - p_i^C = \vartheta$ for all i, a straight line with slope 1 and intercept ϑ:

$$p_i^T = p_i^C + \vartheta \text{ for all } i.$$

Table 1.12 *Data from the multicenter Beta-Blocker Heart Attack Trial (BHAT) comparing propranolol with placebo (Horwitz et al. (1996))*

Trial	Treatment		Control	
i	x_i^T	n_i^T	x_i^C	n_i^C
1	0	49	3	48
2	1	57	7	58
3	1	56	5	57
4	1	42	4	42
5	3	65	10	65
6	3	70	8	71
7	2	65	5	66
8	3	55	7	55
9	3	55	7	56
10	2	44	4	44
11	4	77	8	78
12	2	58	4	59
13	4	70	7	70
14	7	98	12	98
15	3	48	5	48
16	7	52	11	53
17	5	64	7	65
18	3	42	4	42
19	5	47	6	47
20	4	62	5	63
21	5	59	6	60
22	5	59	1	59
23	8	63	3	64
24	7	55	4	55
25	8	75	5	75
26	7	91	6	92
27	8	70	7	70
28	3	44	3	44
29	4	32	4	33
30	10	57	10	58
31	10	125	10	126

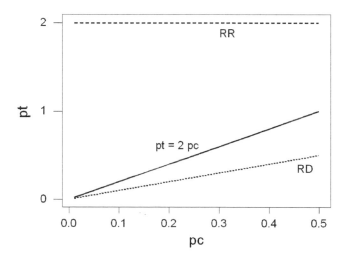

Figure 1.2 *Artificial heterogeneity of effect due to wrong effect measure: p_i^T has been generated using θp_i^C with $\theta = 2$ and $p_i^C = 0.01, 0.02, ..., 0.50$ leading to constant $\theta_i = 2$, but increasing risk difference ϑ_i*

Finally, for the odds ratio the simplest model is $\frac{p_i^T/(1-p_i^T)}{p_i^C/(1-p_i^C)} = \kappa$ for all i, or

$$p_i^T/(1-p_i^T) = p_i^C/(1-p_i^C)\kappa \text{ for all } i,$$

a straight line through the origin with slope κ in the $(p_i^T/(1-p_i^T), p_i^C/(1-p_i^C))'$-plane. Figure 1.3 gives an illustration for the relative risk and the risk difference at hand of the data of Table 1.5. The points (circles) in Figure 1.3 correspond to the observed pairs $(x_i^C/n_i^C, x_i^T/n_i^T)'$, whereas the two straight lines correspond to the relative risk model (dotted) and the risk difference (dashed). These lines were found by regressing $y_i = x_i^T/n_i^T$ on $z_i = x_i^C/n_i^C$, in the case of the relative risk model *without* intercept, and in the case of the risk difference with free intercept and slope equals one. The solid line in Figure 1.3 corresponds to the best line with free intercept and free slope parameter. In the case of the data of Table 1.5, the relative risk model is closer to the best line.

1.6.1 Relative risk line

The least squares estimate of $y_i = \theta z_i$ for $i = 1, ..., k$ is given as

$$\hat{\theta}_{ls} = \frac{\sum_{i=1}^{k} y_i z_i}{\sum_{i=1}^{k} z_i^2} = \frac{\sum_{i=1}^{k}(x_i^T/n_i^T)(x_i^C/n_i^C)}{\sum_{i=1}^{k}(x_i^C/n_i^C)^2}. \tag{1.4}$$

Though this estimate is intuitively appealing it has the disadvantage to weigh each of the centers equally. We prefer to use the Mantel-Haenszel estimate in this setting (Woodward (1999)). Let us write the center-specific relative risk as $\frac{x_i^T/n_i^T}{x_i^C/n_i^C} = \frac{x_i^T n_i^C/(n_i^T+n_i^C)}{x_i^C n_i^T/(n_i^T+n_i^C)}$. Now the Mantel-Haenszel estimate occurs if we take summation before taking ratios:

$$\hat{\theta}_{MH} = \frac{\sum_{i=1}^{k} x_i^T n_i^C/(n_i^T + n_i^C)}{\sum_{i=1}^{k} x_i^C n_i^T/(n_i^T + n_i^C)} \tag{1.5}$$

Note that (1.4) and (1.5) agree if $k = 1$. The Mantel-Haenszel estimate is a weighted estimator $\sum_i w_i \frac{x_i^T n_i^C}{x_i^C n_i^T} / \sum w_i$ with weights $w_i = x_i^C n_i^T/(n_i^T + n_i^C)$. It is popular with epidemiologists since it experiences stability under sparsity such as being less affected by zero events in the centers.

1.6.2 Risk difference line

Similarly, the least squares estimate of $y_i = \vartheta + z_i$ for $i = 1, ..., k$ is given as

$$\hat{\vartheta}_{ls} = \frac{1}{k}\sum_{i=1}^{k} y_i - \frac{1}{k}\sum_{i=1}^{k} z_i = \frac{1}{k}\sum_{i=1}^{k}(x_i^T/n_i^T) - \frac{1}{k}\sum_{i=1}^{k}(x_i^C/n_i^C). \tag{1.6}$$

Again, we prefer to use the Mantel-Haenszel estimate in this setting (Böhning (2000)). Let us write the center-specific risk difference as $x_i^T/n_i^T - x_i^C/n_i^C = \frac{(x_i^T n_i^C - x_i^C n_i^T)/n_i}{(n_i^T n_i^C/n_i)}$, where $n_i = n_i^T + n_i^C$. Now the Mantel-Haenszel estimate occurs if we take summation before taking ratios:

$$\hat{\vartheta}_{MH} = \frac{\sum_{i=1}^{k}(x_i^T n_i^C - x_i^C n_i^T)/n_i}{\sum_{i=1}^{k}(n_i^T n_i^C)/n_i} \tag{1.7}$$

Note that (1.6) and (1.7) agree if $k = 1$. The Mantel-Haenszel estimate is again a weighted estimator $\sum_i w_i(x_i^T/n_i^T - x_i^C/n_i^C)/ \sum w_i$ with weights $w_i = n_i^C n_i^T/n_i$.

1.6.3 Odds ratio line

The least squares estimate of $y_i = \kappa z_i$ for $i = 1, ..., k$ is given as

$$\hat{\kappa}_{ls} = \frac{\sum_{i=1}^k y_i z_i}{\sum_{i=1}^k z_i^2} = \frac{\sum_{i=1}^k (\frac{x_i^T/n_i^T}{1-x_i^T/n_i^T})(\frac{x_i^C/n_i^C}{1-x_i^C/n_i^C})}{\sum_{i=1}^k (\frac{x_i^C/n_i^C}{1-x_i^C/n_i^C})^2}. \tag{1.8}$$

Note that $y_i = \frac{x_i^T/n_i^T}{1-x_i^T/n_i^T}$, the odds in the treatment arm, and $z_i = \frac{x_i^C/n_i^C}{1-x_i^C/n_i^C}$ in the control arm. In other words, we are considering now the straight line through the origin in the odds plane. Though this estimate is intuitively appealing it has several disadvantages including weighing each center equally. We prefer to use the Mantel-Haenszel estimate in this setting (Woodward (1999)). Let us write the center-specific odds ratio as $\frac{x_i^T/(n_i^T-x_i^T)}{x_i^C/(n_i^C-x_i^C)} = \frac{x_i^T(n_i^C-x_i^C)/(n_i^T+n_i^C)}{x_i^C(n_i^T-x_i^T)/(n_i^T+n_i^C)}$. Now the Mantel-Haenszel estimate occurs if we take summation before taking ratios:

$$\hat{\kappa}_{MH} = \frac{\sum_{i=1}^k x_i^T(n_i^C - x_i^C)/(n_i^T + n_i^C)}{\sum_{i=1}^k x_i^C(n_i^T - x_i^T)/(n_i^T + n_i^C)} \tag{1.9}$$

Note that (1.8) and (1.9) agree if $k = 1$. The Mantel-Haenszel estimate is a weighted estimator $\sum_i w_i \frac{x_i^T(n_i^C-x_i^C)/(n_i^T+n_i^C)}{x_i^C(n_i^T-x_i^T)/(n_i^T+n_i^C)} / \sum w_i$ with weights $w_i = x_i^C(n_i^T - x_i^T)/(n_i^T + n_i^C)$.

1.6.4 Assessment of effect measure for the homogeneity case

We now consider a more formal assessment for the three models of homogeneity: risk difference, risk ratio, and odds ratio. Consider the Poisson log-likelihood

$$\sum_{i=1}^k -n_i^T p_i^T + x_i^T \log(p_i^T) - n_i^C p_i^C + x_i^C \log(p_i^C). \tag{1.10}$$

If p_i^C is estimated by x_i^C/n_i^C, then the three likelihoods will differ only with respect to

$$\sum_{i=1}^k -n_i^T \hat{p}_i^T + x_i^T \log(\hat{p}_i^T) \tag{1.11}$$

where $\hat{p}_i^T = \hat{p}_i^C + \hat{\vartheta}_{MH}$, $\hat{p}_i^T = \hat{p}_i^C \hat{\theta}_{MH}$, and $\frac{\hat{p}_i^T}{1-p_i^T} = \frac{\hat{p}_i^C}{1-p_i^C} \hat{\kappa}_{MH}$, respectively.

In the following we will look at various MAIPDs and their associated log-likelihoods (1.11). Table 1.13 shows the corresponding log-likelihoods. Often the measure of relative risk provides the largest log-likelihood.

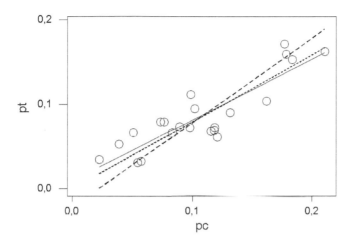

Figure 1.3 *Relative risk (dotted line) and risk difference model (dashed line) in comparison to the best (regression) line for data of Table 1.5*

Table 1.13 *Assessment of effect measure for the homogeneity case by means of the log-likelihood*

MAIPD	Log-likelihood		
Table	$\hat{\theta}_{MH}$	$\hat{\vartheta}_{MH}$	$\hat{\kappa}_{MH}$
1.5	-2883.0950	-2907.7720	-2883.6310
1.6	-13329.4600	-13348.0300	-13329.7300
1.7	-139.0275	-137.6152	-138.8530
1.8	-77.0129	-78.7789	-78.0682
1.9	-1264.721	-1222.849	-1252.038
1.10	-3666.9060	-3668.8850	-3666.9230
1.11	-54.3502	-51.0569	-51.797
1.12	-648.6063	-634.8513	-647.4781

Table 1.14 provides the associated Mantel-Haenszel estimators for the MAIPDs presented in Tables 1.5 to 1.12. As an additional analysis tool we consider the

Table 1.14 *The results of MH-estimator*

MAIPD	MH-estimator		
Table	$\hat{\theta}_{MH}$	$\hat{\vartheta}_{MH}$	$\hat{\kappa}_{MH}$
1.5	0.7908	-0.0209	0.7702
1.6	0.9708	-0.0007	0.9695
1.7	1.7345	.0281	1.7893
1.8	0.8991	-0.0572	0.7883
1.9	2.1238	0.1531	2.8225
1.10	1.1079	0.0007	1.1088
1.11	10.0428	0.4160	20.0626
1.12	1.3506	0.0254	1.3903

χ^2-test of homogeneity* defined as

$$\chi^2_{hom} = \sum_i \frac{(\hat{\theta}_i - \hat{\theta}_{MH})^2}{\widehat{var}(\theta_i)},$$

here for the risk ratio and similarly for the other effect measures. We emphasize in Table 1.15 on the comparison of relative risk and risk difference. The χ^2-test statistic is here used in a descriptive statistical sense, not as an inferential statistical tool, since its distributional properties are in doubt (Hartung et al. (2003), Hartung and Knapp (2003), Knapp et al. (2006), Jackson (2006)). As Table 1.15 shows in most cases the relative risk provides the better fitting model. These considerations underline the importance of the choice of the effect measure and provide empirical evidence for the relative risk as an appropriate choice as an effect measure.

---

* The χ^2-test of homogeneity measures the degree of heterogeneity (variation between studies) of the study-specific relative risks. The statistical package STATA (Tables for Epidemiologists) (StataCorp. (2005)) was used to calculate the values of the test statistics. For the MAIPD of Table 1.10, there is only a risk ratio, so that a comparison of different effect measures was not possible.

Table 1.15 χ^2_{hom}-test of homogeneity (M-H)

MAIPD	RR		OR	
Table	χ^2_{hom}	p-value	χ^2_{hom}	p-value
1.5	23.041	0.3418	76.671	0.0000
1.6	76.380	0.0000	76.671	0.0000
1.7	1.580	0.9037	1.515	0.9113
1.8	11.682	0.8632	4.671	0.9946
1.9	40.783	0.0040	57.493	0.0000
1.10	-	-	-	-
1.11	2.641	0.6195	2.821	0.4201
1.12	25.508	0.6517	25.899	0.6309

CHAPTER 2

The basic model

Modeling effects in meta-analysis of trials is of interest and has been under investigation for quite some time. For an overview see Agresti and Hartzel (2000). In this section we would like to follow a strict likelihood approach. In the likelihood theory, when dealing with nuisance parameters, one approach (see Murphy and Van der Vaart (2000) or Pawitan (2001)) uses the profile likelihood method to eliminate the nuisance parameters. We will show in this section that this approach for a MAIPD leads to a very simple and easy to handle profile likelihood which can be used in more elaborate developments later on.

Let x_i^T and x_i^C again denote the number of events in treatment and control arms, respectively, with n_i^T the person-time in the treatment arm and n_i^C denoting the person-time in the control arm. Let the number of trials be k, so that $i = 1, ..., k$. Also, let p_i^T and p_i^C denote the risk of an event in the treatment and control arm, respectively. Typically, we will be interested in effect measures of treatment like the *risk ratio* $\theta_i = p_i^T / p_i^C$.

2.1 Likelihood

We are interested in the inference on $\theta_i = p_i^T / p_i^C$, the ratio of the two event probabilities p_i^T for the treatment arm and p_i^C for the control arm. In contrast to the single study settings, we have to investigate the variation of the risk ratio between studies in the case of MAIPD. If *homogeneity* of effect can be established, the results are more *supportive* of the effect. If *heterogeneity* is present, an appropriate modeling is required and sources for its occurrence should be investigated.

For each trial and for each arm there is a Poisson likelihood, so that for the i-th trial the contribution to the likelihood of the treatment arm is

$$\exp(-n_i^T p_i^T)(n_i^T p_i^T)^{x_i^T} / x_i^T! \tag{2.1}$$

and for the i-th trial the contribution to the likelihood of the control arm is

$$exp(-n_i^C p_i^C)(n_i^C p_i^C)^{x_i^C} / x_i^C! \tag{2.2}$$

so that the product likelihood over *all* trials becomes

$$\prod_{i=1}^{k} \left(\exp(-n_i^T p_i^T)(n_i^T p_i^T)^{x_i^T} / x_i^T! \times \exp(-n_i^C p_i^C)(n_i^C p_i^C)^{x_i^C} / x_i^C! \right) \qquad (2.3)$$

and the *log-likelihood* takes the form* (ignoring the only data-dependent terms)

$$\sum_{i=1}^{k} \left\{ -n_i^T p_i^T + x_i^T \log(p_i^T) - n_i^C p_i^C + x_i^C \log(p_i^C) \right\}. \qquad (2.4)$$

2.2 Estimation of relative risk in meta-analytic studies using the profile likelihood

Rewrite p_i^T as $p_i^C \theta_i$ and (2.4) becomes

$$\sum_{i=1}^{k} \left\{ -n_i^T p_i^C \theta_i + x_i^T \log(p_i^C \theta_i) \quad n_i^C p_i^C + x_i^C log(p_i^C) \right\} \qquad (2.5)$$

Note that in the log-likelihood (2.5), there occurs two kinds of parameters: the effect measuring parameter θ_i or *the parameter of interest*; and the baseline parameter p_i^C or *the nuisance parameter*. In general, let the log-likelihood $L(\mathbf{p}, \mathbf{q})$ depend on a vector $\mathbf{p}$ of parameters of interest and a vector $\mathbf{q}$ of nuisance parameters. Let $L(\mathbf{q}|\mathbf{p}) = L(\mathbf{p}, \mathbf{q})$ be the log-likelihood for arbitrary but fixed $\mathbf{p}$, and let $\mathbf{q_p}$ be such that $L(\mathbf{q_p}|\mathbf{p}) \geq L(\mathbf{q}|\mathbf{p})$ for all $\mathbf{q}$, then

$$L^*(\mathbf{p}) = L(\mathbf{q_p}|\mathbf{p}) \qquad (2.6)$$

is called the *profile log-likelihood*. Note that the profile log-likelihood is now depending only on the parameters of interest and, thus, the method of profile log-likelihood can be viewed as a method to deal with nuisance parameters.

Let us determine the *profile log-likelihood* on the basis of (2.5), which we now consider as a function of $\mathbf{p}^C$ for arbitrary, but fixed $\theta = (\theta_1, ..., \theta_k)'$:

$$L(\mathbf{p}^C|\theta) = \sum_{i=1}^{k} \left\{ -n_i^T p_i^C \theta_i + x_i^T log(\theta_i) - n_i^C p_i^C + (x_i^C + x_i^T)log(p_i^C) \right\} \qquad (2.7)$$

$$= \sum_{i=1}^{k} \left\{ -(n_i^C + n_i^T \theta_i)p_i^C + x_i^T log(\theta_i) + (x_i^C + x_i^T)log(p_i^C) \right\}.$$

To determine $\mathbf{p}_\theta^C$ (which maximizes (2.7)) we calculate the partial derivatives

$$\frac{\partial}{\partial p_j^C} L(\mathbf{p}^C|\theta) = -(n_j^C + n_j^T \theta_j) + (x_j^T + x_j^T)/p_j^C \qquad (2.8)$$

---

* `log` we always denote the natural logarithm, e.g., with respect to base e

which can be readily solved for p_j^C as

$$p_{j\theta}^C = \frac{x_j^C + x_j^T}{n_j^C + \theta_j n_j^T}.$$ (2.9)

Inserting (2.9) into (2.7) leads to

$$\sum_{i=1}^{k} \left\{ -(n_i^C + \theta_i n_i^T)(\frac{x_i^C + x_i^T}{n_i^C + \theta_i n_i^T}) + x_i^T \log(\theta_i) + (x_i^C + x_i^T) \log(\frac{x_i^C + x_i^T}{n_i^C + \theta_i n_i^T}) \right\}$$ (2.10)

which simplifies to (if we only consider parameter-dependent terms)

$$L^*(\theta) = \sum_{i=1}^{k} \left\{ x_i^T \log(\theta_i) - (x_i^C + x_i^T) \log(n_i^C + \theta_i n_i^T) \right\}.$$ (2.11)

The above expression $L^*(\theta)$ is the *profile log-likelihood* for the risk ratio and all inference will be based upon this log-likelihood.

2.3 The profile likelihood under effect homogeneity

This section investigates the profile likelihood method for the situation of homogeneity and shows its connection to the Mantel-Haenszel approach, and, thus, is of interest in itself. To illustrate the simplicity and usefulness of the profile method consider the situation of *homogeneity* of effect: $\theta_1 = \theta_2 = ... = \theta_k = \theta$. Then, taking the derivative of (2.11) w.r.t. θ, yields

$$\sum_{i=1}^{k} \left(x_i^T / \theta - (x_i^C + x_i^T) n_i^T / (n_i^C + \theta n_i^T) \right) = 0,$$ (2.12)

or, equivalently

$$\sum_{i=1}^{k} \frac{w_i(\theta)}{\theta} \left(x_i^T n_i^C - x_i^C n_i^T \theta) \right) = 0,$$ (2.13)

where $w_i(\theta) = 1/(n_i^C + \theta n_i^T)$. Equation (2.13) is an implicit feature of the maximum profile likelihood estimator of relative risk which can be further written as

$$\theta = \frac{\sum_{i=1}^{k} w_i(\theta) x_i^T n_i^C}{\sum_{i=1}^{k} w_i(\theta) x_i^C n_i^T},$$ (2.14)

which can be used to iteratively construct the maximum likelihood estimator.

2.3.1 Score of profile likelihood

Let us consider now the profile likelihood in more detail. The profile log-likelihood (2.11) becomes

$$L^*(\theta) = \sum_{i=1}^{k} x_i^T log(\theta) - \sum_{i=1}^{k} (x_i^C + x_i^T) log(n_i^C + \theta n_i^T). \tag{2.15}$$

The log-likelihood (2.15) has no closed form maximum likelihood solution. The first derivative is

$$L^{*\prime}(\theta) = \frac{\sum_{i=1}^{k} x_i^T}{\theta} - \sum_{i=1}^{k} \frac{(x_i^C + x_i^T) n_i^T}{n_i^C + \theta n_i^T}. \tag{2.16}$$

We first show the existence of a solution for the score equation of the profile maximum likelihood estimator, then its uniqueness, and, finally, that it is indeed a maximum.

2.3.2 Existence

We write (2.16) as $f(\theta)/\theta$ where $f(\theta) = \sum_{i=1}^{k} x_i^T - \sum_{i=1}^{k} \frac{(x_i^C + x_i^T)\theta n_i^T}{n_i^C + \theta n_i^T}$ since, for positive θ, the zeros of $f(\theta)/\theta$ are identical to the zeros of $f(\theta)$. Now, $f(\theta)$ approaches $\sum_{i=1}^{k} x_i^T$ if θ approaches 0 and $f(\theta)$ approaches $\sum_{i=1}^{k} x_i^T - \sum_{i=1}^{k} (x_i^C + x_i^T) = -\sum_{i=1}^{k} x_i^C$ if θ goes to ∞. In other words, $f(\theta)$ changes its sign, from positive to negative if θ moves from 0 to ∞. Therefore, (2.16) must have at least one zero. If we equate (2.16) to zero, then we have the *score equation* for the profile maximum likelihood estimator.

2.3.3 Uniqueness of solution of score equation

The score (2.16) equals 0, if and only if $f(\theta) = 0$. Now,

$$f'(\theta) = -\sum_{i=1}^{k} \frac{(x_i^C + x_i^T) n_i^T (n_i^C + n_i^T \theta) - n_i^T (x_i^C + x_i^T)\theta n_i^T}{(n_i^C + \theta n_i^T)^2} \tag{2.17}$$

$$= -\sum_{i=1}^{k} \frac{(x_i^C + x_i^T) n_i^T n_i^C}{(n_i^C + \theta n_i^T)^2} < 0.$$

for all θ. Therefore, $f(\theta)$ is strictly monotone decreasing in θ and can have at most one zero.

2.3.4 Solution of score equation is maximum

A solution of (2.16) is not necessarily a maximum. Let $\hat{\theta}$ be any solution of (2.16). We show that the profile log-likelihood is locally concave at $\hat{\theta}$ by

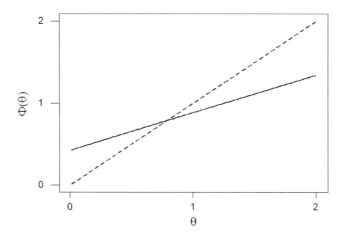

Figure 2.1 *Graph of $\Phi(\theta)$ for the data of Table 1.5*

proving that its second derivative is nonpositive at $\hat{\theta}$. Consider the second derivative

$$L^{*''}(\theta) = -\frac{\sum_{i=1}^{k} x_i^T}{\theta^2} + \sum_{i=1}^{k} \frac{(x_i^C + x_i^T)(n_i^T)^2}{(n_i^C + \theta n_i^T)^2} \tag{2.18}$$

$$= -\frac{\sum_{i=1}^{k} x_i^T}{\theta^2} + \frac{1}{\theta^2} \sum_{i=1}^{k} \frac{(x_i^C + x_i^T)(\theta n_i^T)^2}{(n_i^C + \theta n_i^T)^2}.$$

Furthermore, at the solution of the score equation we have that $\frac{\sum_{i=1}^{k} x_i^T}{\hat{\theta}} = \sum_{i=1}^{k} \frac{(x_i^C + x_i^T)n_i^T}{n_i^C + \hat{\theta} n_i^T}$, so that (2.18) at $\hat{\theta}$ becomes

$$L^{*''}(\hat{\theta}) = -\frac{1}{\hat{\theta}^2} \sum_{i=1}^{k} \frac{(x_i^C + x_i^T)(\hat{\theta} n_i^T)}{(n_i^C + \hat{\theta} n_i^T)} + \frac{1}{\hat{\theta}^2} \sum_{i=1}^{k} \frac{(x_i^C + x_i^T)(\hat{\theta} n_i^T)^2}{(n_i^C + \hat{\theta} n_i^T)^2} \tag{2.19}$$

which can be written as

$$L^{*''}(\hat{\theta}) = -\frac{1}{\hat{\theta}^2} \sum_{i=1}^{k} (x_i^C + x_i^T)\alpha_i(1 - \alpha_i) \le 0 \tag{2.20}$$

since $\alpha_i = \frac{\hat{\theta} n_i^T}{n_i^C + \hat{\theta} n_i^T} < 1$ for all i.

2.4 Reliable construction of the profile MLE

The score equation leads to $f(\theta) = \sum_{i=1}^k x_i^T - \sum_{i=1}^k \frac{(x_i^C + x_i^T)\theta n_i^T}{n_i^C + \theta n_i^T} = 0$ or

$$\sum_{i=1}^k x_i^T = \sum_{i=1}^k \frac{(x_i^C + x_i^T)\theta n_i^T}{n_i^C + \theta n_i^T} \tag{2.21}$$

which can be equivalently written as

$$\theta = \frac{\sum_{i=1}^k x_i^T}{\sum_{i=1}^k \frac{(x_i^C + x_i^T)n_i^T}{n_i^C + \theta n_i^T}} =: \Phi(\theta) \tag{2.22}$$

where we take the right-hand side of (2.22) as a definition of the mapping $\Phi(\theta)$. Evidently, the score equation has been equivalently transferred to a fixed point equation $\Phi(\theta) - \theta$. A graph of this fixed point mapping is provided in Figure 2.1 for the data of Table 1.5.

The benefit of this equivalence can be seen in the fact that the fixed point mapping can be used in a constructive way to generate the profile maximum likelihood estimator. Start with some initial value $\theta^{(0)}$ and use Φ to yield $\theta^{(1)} = \Phi(\theta^{(0)})$, in general, $\theta^{(n)} = \Phi(\theta^{(n-1)})$ for $n = 1, 2, 3, \dots$. The question arises under which conditions this sequence $(\theta^{(n)})$ will converge to the profile maximum likelihood estimator $(\hat{\theta}_{PMLE})$? To answer this question with a clear *yes* we need a result from fixed point theory.

Lemma 4.1 Let $\Phi(\theta)$ be a real-value mapping. If there exist values θ^L and θ^U such that

1. $\theta^L < \theta^U$
2. $\Phi(\theta^L) \geq \theta^L$ and $\Phi(\theta^U) \leq \theta^U$
3. Φ is monotone, e.g., $\theta_1 \leq \theta_2$ implies $\Phi(\theta_1) \leq \Phi(\theta_2)$

then any sequence (θ^n) constructed by $\theta^{(n)} = \Phi(\theta^{(n-1)})$ with $\theta^L \leq 0^{(0)} \leq \theta^U$ converges to a fixed point of Φ.

We now verify that Φ defined in (2.22) fulfills the conditions of Lemma 4.1. To verify monotonicity we show that the first derivative of Φ is positive. Indeed,

$$\Phi'(\theta) = \frac{\sum_{i=1}^k x_i^T \sum_{i=1}^k \frac{(x_i^C + x_i^T)(n_i^T)^2}{(n_i^C + \theta n_i^T)^2}}{(\sum_{i=1}^k \frac{(x_i^C + x_i^T)n_i^T}{n_i^C + \theta n_i^T})^2} > 0$$

for all $\theta \geq 0$. This proves monotonicity. To find θ^L and θ^U note that $\Phi(0) = \frac{\sum_{i=1}^k x_i^T}{\sum_{i=1}^k \frac{(x_i^C + x_i^T)n_i^T}{n_i^C}} > 0$. In other words, we can take $\theta^L = 0$. To prove the exis-

tence of a θ^U, we will show that $\lim_{\theta \to \infty} \Phi(\theta)/\theta < 1$. Now,

$$\Phi(\theta)/\theta = \frac{\sum_{i=1}^k x_i^T}{\sum_{i=1}^k \frac{(x_i^C + x_i^T)\theta n_i^T}{n_i^C + \theta n_i^T}} = \tag{2.23}$$

$$= \frac{\sum_{i=1}^{k} x_i^T}{\sum_{i=1}^{k} \frac{(x_i^C + x_i^T)}{n_i^C / \theta n_i^T + 1}} \rightarrow_{\theta \to \infty} \frac{\sum_{i=1}^{k} x_i^T}{\sum_{i=1}^{k} (x_i^C + x_i^T)} < 1$$

which is the desired result. Since θ^L and θ^U correspond to the left and right endpoints of $(0, \infty)$, so we have that the fixed point procedure is guaranteed to converge from any initial value in $(0, \infty)$.

2.4.1 Numerical illustration

We illustrate this method with the data we have discussed in Chapter 1. Using the data from Table 1.5 we can calculate the result in Table 2.1 which illustrates the convergent behavior of the fixed point sequence. The sequence has been started at $\theta^{(0)} = 1$ corresponding to the situation of no effect. The sequence stopped when $|\theta^{(n-1)} - \theta^{(n)}| \leq 0.00001$. Thirteen iterations were required to reach this stop rule. To measure the *rate of convergence* it is customary to look at the ratios of differences $r_n = \frac{|\theta^{(n+1)} - \theta^{(n)}|}{|\theta^{(n-1)} - \theta^{(n)}|}$. This can be constructed from column 4 of Table 2.1 and is approximately $r_n = 0.4595$ for all n. Evidently, the ratio r_n is constant for all n and this case is called *linear convergence*. In theoretical terms the linear convergence rate is provided by the first derivative of the fixed point mapping at θ_{PMLE}. Clearly, the smaller this value is in absolute terms, the faster is the convergence. The best would be a situation in which $\Phi'(\theta_{PMLE}) \approx 0$. This corresponds to a mapping, which is almost horizontally tangent at the fixed point. Consider Figure 2.1: the slope of the tangent at the fixed point is not close to zero at all.

However, things can become worse. Consider Figure 2.2: the slope of the tangent at the fixed point is closer to one than in Figure 2.1. Consequently, the convergence rate is bad, and the sequence needs considerable time to reach the fixed point.

2.5 A fast converging sequence

Let us consider again (2.21)

$$\sum_{i=1}^{k} x_i^T = \sum_{i=1}^{k} \frac{(x_i^C + x_i^T) \theta n_i^T}{n_i^C + \theta n_i^T}$$

which we write as

$$\sum_{i=1}^{k} \frac{x_i^T (n_i^C + \theta n_i^T)}{n_i^C + \theta n_i^T} = \sum_{i=1}^{k} \frac{(x_i^C + x_i^T) \theta n_i^T}{n_i^C + \theta n_i^T} \tag{2.24}$$

or equivalently,

$$\sum_{i=1}^{k} x_i^T n_i^C w_i(\theta) = \sum_{i=1}^{k} x_i^C n_i^T w_i(\theta) \theta$$

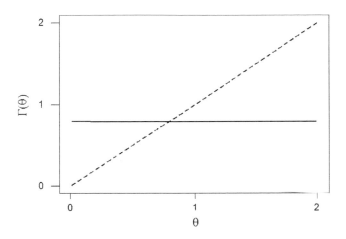

Figure 2.2 *Graph of* $\Phi(\theta)$ *for the data of table 1.11*

with $w_i(\theta) = 1/(n_i^C + \theta n_i^T)$. (2.24) can be solved for θ such that

$$\theta = \frac{\sum_{i=1}^{k} x_i^T n_i^C w_i(\theta)}{\sum_{i=1}^{k} x_i^C n_i^T w_i(\theta)} := \Gamma(\theta). \qquad (2.25)$$

(2.25) has the feature that for $\theta = 1$ the *Mantel-Haenszel estimate* of relative risk occurs:

$$\hat{\theta}_{MH} = \frac{\sum_{i=1}^{k} x_i^T n_i^C w_i(1)}{\sum_{i=1}^{k} x_i^C n_i^T w_i(1)} = \frac{\sum_{i=1}^{k} x_i^T n_i^C/(n_i^C + n_i^T)}{\sum_{i=1}^{k} x_i^C n_i^T/(n_i^C + n_i^T)}.$$

Also the fixed point iteration based upon $\Gamma(\theta)$ provides a fast converging sequence $\theta^{(n)} = \Gamma(\theta^{(n-1)})$. In Figure 2.3, the slope of $\Gamma(\theta)$ at the fixed point is horizontal, providing a faster converging sequence. This sequence is called *quadratically convergent* to $\hat{\theta}$ if $\frac{|\theta^{(n+1)} - \hat{\theta}|}{|\theta^{(n)} - \hat{\theta}|^2} \leq C$ for some constant C and all n. It follows that a sequence based upon the fixed point mapping $\Gamma(\theta)$ is quadratically convergent if $\Gamma'(\theta)|_{\hat{\theta}} = 0$ [†]. It appears that this is the case for the data of Table (1.5) as demonstrated in Figure 2.3. Consider the data of Table 1.11 where the sequence based upon Φ has been converging very slowly. For those data the fixed point mapping Γ is given in Figure 2.4. Again, convergent

[†] This follows from a second order Taylor expansion of $\Gamma(\theta)$ around $\hat{\theta}$: $\Gamma(\theta^{(n)}) \approx \Gamma(\hat{\theta}) + \Gamma'(\hat{\theta})(\theta^{(n)} - \hat{\theta}) + \Gamma''(\hat{\theta})(\theta^{(n)} - \hat{\theta})^2/2 = \Gamma(\hat{\theta}) + \Gamma''(\hat{\theta})(\theta^{(n)} - \hat{\theta})^2/2$. We assume that $\theta^{(n+1)} - \hat{\theta} \leq (\theta^{(n)} - \hat{\theta})^2 \times C$ which serves as boundedness of the second derivative.

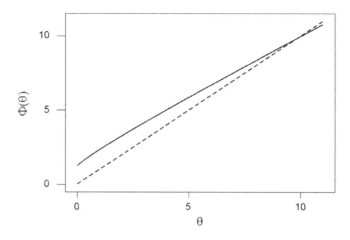

Figure 2.3 *Graph of* $\Gamma(\theta)$ *for the data of Table 1.5*

behavior appears to be quadratic (a general proof is still outstanding), and it can be expected that the iteration will only need a few steps. Indeed, for this case iteration based upon Γ stops only after 5 steps, whereas the iteration based upon Φ stops after 60 steps.

2.5.1 Comparing two fixed point iterations

A more complete comparison of the fixed point mappings of Φ and Γ is provided in Table 2.4. Here, both iterative schemes are compared with respect to the question of which of the two uses fewer steps to reach the stop rule that two consecutive iterations are smaller than 0.00001. In all cases considered in Table 2.4 the fixed point procedure Γ using the Mantel-Haenszel weights requires considerably fewer steps to complete computation in comparison to the fixed point procedure Φ. This makes the iteration (2.25) more suitable for practical use. On the other hand, there is no proof of convergence for the iteration (2.25). In fact, the fixed point mapping is *not* monotone as Figure 2.4 shows. Although nonconvergence of a sequence based upon the iteration (2.25) was never observed in practice, more analysis and experience with this mapping should be collected. However, we would like to mention one possible explanation for the excellent convergent behavior of the sequence based upon the iteration (2.25). In the balanced case, e.g., $n_i^T = n_i^C$ for all i, the iteration based on (2.25) converges in *one* step. This also means that in this case there exists a closed form solution for the profile maximum likelihood estimator. Since often multicenter trials are balanced or nearly balanced, only a few steps are required with this iteration.

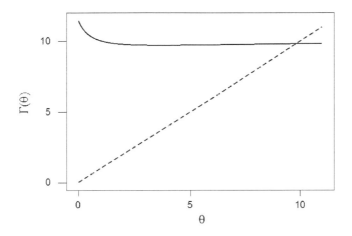

Figure 2.4 *Graph of* $\Gamma(\theta)$ *for the data of Table 1.11*

If iteration is started with $\theta = 1$, the first iteration using (2.26) leads to the well-known *Mantel-Haenszel estimator* of the risk ratio:

$$\theta = \frac{\sum_{i=1}^{k} w_i(1)x_i^T n_i^C}{\sum_{i=1}^{k} w_i(1)x_i^C n_i^T} = \frac{\sum_{i=1}^{k} x_i^T n_i^C / n_i}{\sum_{i=1}^{k} x_i^C n_i^T / n_i}, \tag{2.26}$$

where $n_i = n_i^T + n_i^C$ is the total person-time of the i-th center. For the time being, we remain with the Mantel-Haenszel estimator (MHE) $\hat{\theta}_{MH} = \frac{\sum_{i=1}^{k} x_i^T n_i^C / n_i}{\sum_{i=1}^{k} x_i^C n_i^T / n_i}$ and compare it with the PMLE under effect homogeneity. Clearly, if the trial is completely balanced $n_i^T = n_i^C$ for all centers i, then the parameter-dependent weights cancel out, and PMLE and MHE are identical. Typically, MHE and PMLE are close and the loss of efficiency in using the MHE is not high. However, the exception is the situation of sparsity. Simulation studies provide some evidence that in this case there is considerable loss of efficiency, in particular, when θ is bounded away from 1. To demonstrate this finding a simulation experiment was conducted. Since differences between PMLE and MHE can only be expected for highly unbalanced and sparse multicenter studies, n_i^T and n_i^C were generated from Poisson distributions with Poisson parameters 3 and 6, where the arm allocation of the Poisson parameters were random to guarantee that the trial is unbalanced. The baseline parameter p_i^C

was chosen from a uniform distribution with parameters [0.1,0.3]. The risk ratio parameter was kept fixed for each simulation (replication size 10,000) and risk ratio values which ranged from 0.00001 to 3.3333 were considered. The number of centers k was chosen to be 5. The results for the two estimators are provided in Figure 2.5 for the bias and in Figure 2.6 for the variance indicating a superior behavior of the PMLE with respect to both criteria.

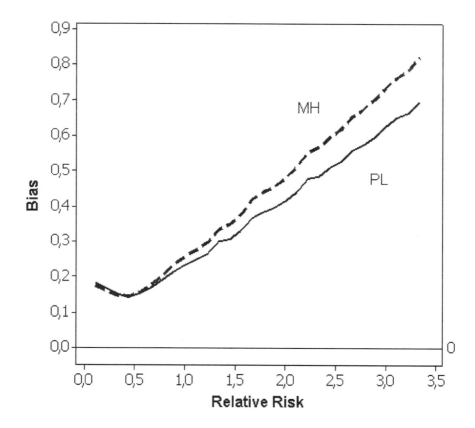

Figure 2.5 *A comparison of the profile maximum likelihood estimator (PL) and the Mantel-Haenszel estimator (MH) for the sparse multicenter trial with respect to bias based upon a simulation*

2.6 Inference under effect homogeneity

2.6.1 Variance estimate of the PMLE

Another beneficial aspect of the profile likelihood method lies in the fact that it easily provides an estimate of the variance of the PMLE. We will use a standard result from likelihood theory (see, for example Le (1992), p. 72–73)

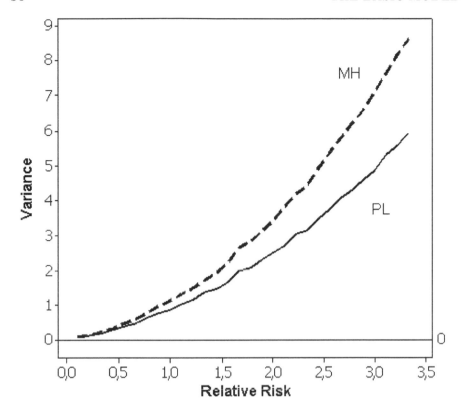

Figure 2.6 *A comparison of the profile maximum likelihood estimator (PL) and the Mantel-Haenszel estimator (MH) for the sparse multicenter trial with respect to variance based upon a simulation*

that the variance of the maximum likelihood estimate can be approximated by the negative inverse of the second derivative of the log-likelihood function which is evaluated at the maximum likelihood estimate. We apply this result to the profile likelihood situation. In our case we have that

$$L^{*\prime\prime}(\theta) = \{[\sum_{i=1}^{k} x_i^T]log(\theta) - \sum_{i=1}^{k}(x_i^C + x_i^T)log(n_i^C + \theta n_i^T)\} \qquad (2.27)$$

$$= -\frac{\sum_{i=1}^{k} x_i^T}{\theta^2} + \sum_{i=1}^{k}\frac{(x_i^C + x_i^T)(n_i^T)^2}{(n_i^C + \theta n_i^T)^2}.$$

It was shown in (2.20) that $L^{*\prime\prime}(\hat{\theta}) < 0$ at the profile maximum likelihood estimate $\hat{\theta} = \hat{\theta}_{PMLE}$. This will guarantee that the estimated variance

$$\widehat{var}(\hat{\theta}) = -1/L^{*\prime\prime}(\hat{\theta}) = \{\frac{\sum_{i=1}^{k} x_i^T}{\hat{\theta}^2} - \sum_{i=1}^{k} \frac{(x_i^C + x_i^T)(n_i^T)^2}{(n_i^C + \hat{\theta}n_i^T)^2}\}^{-1} \qquad (2.28)$$

is positive at the PMLE. (2.28) was further simplified in (2.20) to

$$\widehat{var}(\hat{\theta}) = \hat{\theta}^2 \{\sum_{i=1}^{k} x_i \alpha_i(\hat{\theta})(1 - \alpha_i(\hat{\theta}))\}^{-1} \qquad (2.29)$$

with $x_i = x_i^C + x_i^T$ and $\alpha_i(\hat{\theta}) = \frac{\hat{\theta}n_i^T}{n_i^C + \hat{\theta}n_i^T}$.

2.6.2 Variance estimate of the Mantel-Haenszel estimate

The Mantel-Haenszel estimate of the common relative risk is given as $\hat{\theta}_{MH} = \frac{\sum_i x_i^T n_i^C / n_i}{\sum_i x_i^C n_i^C / n_i}$ with $n_i = n_i^C + n_i^T$. Although the formula for the Mantel-Haenszel relative risk estimate is quite elementary, a widely accepted expression for its variance has been only given recently (Greenland and Robins (1985), see also Woodward (1999)). We have that

$$\widehat{var}(log(\hat{\theta}_{MH})) = \frac{\sum_i (n_i^T n_i^C x_i - x_i^T x_i^C n_i)/(n_i)^2}{(\sum_i x_i^T n_i^C / n_i)(\sum_i x_i^C n_i^T / n_i)} \qquad (2.30)$$

where $x_i = x_i^C + x_i^T$ as before.

2.6.3 Comparing the two variance approximations

Since the Mantel-Haenszel estimator is very popular it appears valuable to compare both variance approximations, namely based upon the profile likelihood (2.29) and the one based upon the Greenland-Robins formula (2.34). Here, one problem occurs, namely that the variance (2.29) is given on the relative risk scale whereas (2.34) is given on the log-relative risk scale. One could use the δ-method, so that $\widehat{var}(\hat{\theta}_{MH}) \approx (e^{\hat{\theta}_{MH}})^2 \widehat{var}(log(\hat{\theta}_{MH}))$, though we prefer to give a more direct comparison.

Let us write the profile log-likelihood (2.11) using $\phi = log(\theta)$

$$L^*(\phi) = [\sum_{i=1}^{k} x_i^T]\phi - \sum_{i=1}^{k}(x_i^C + x_i^T)log(n_i^C + e^{\phi}n_i^T), \qquad (2.31)$$

with a second derivative

$$L^{*\prime\prime}(\phi) = -\sum_{i=1}^{k} \frac{x_i n_i^T n_i^C e^{\phi}}{(n_i^C + e^{\phi}n_i^T)^2} = -\sum_{i=1}^{k} \frac{x_i n_i^T n_i^C \theta}{(n_i^C + \theta n_i^T)^2} = -\sum_{i=1}^{k} x_i \alpha_i(1 - \alpha_i),$$

$$\qquad (2.32)$$

so that an estimate of the variance of the PMLE of ϕ is provided as

$$\widehat{var(\hat\phi)} = \widehat{var(\log\hat\theta)} = \left(\sum_{i=1}^{k} x_i\alpha_i(1-\alpha_i)\right)^{-1}, \qquad (2.33)$$

where $\alpha_i = \frac{\hat\theta n_i^T}{n_i^C + \hat\theta n_i^T}$. The Mantel-Haenszel estimate of the common relative risk is given as $\hat\theta_{MH} = \frac{\sum_i x_i^T n_i^C/n_i}{\sum_i x_i^C n_i^T/n_i}$ with $n_i = n_i^C + n_i^T$ (Greenland and Robins (1985), see also Woodward (1999)). We have the Greenland-Robins formula for the variance of the logarithm of the Mantel-Haenszel estimator as

$$\widehat{var}(log(\hat\theta_{MH})) = \frac{\sum_i(n_i^T n_i^C x_i - x_i^T x_i^C n_i)/(n_i)^2}{(\sum_i x_i^T n_i^C/n_i)(\sum_i x_i^C n_i^T/n_i)} \qquad (2.34)$$

where $x_i = x_i^C + x_i^T$ as before. We note that (2.34) has been developed for the situation of identical person-times in the centers reflecting a binomial sampling plan. Breslow (1984) provided a robust variance formula for the situation of person-specific observation times.

Typically, not only MHE and PMLE, but also the variances of the MHE (2.34) and of the PMLE (2.33) are close. However, the exception is the situation of sparsity. Simulation studies provide some evidence that in this case the variance estimator (2.33) is behaving better than in (2.34). To demonstrate this fact the following simulation experiment was conducted. In the balanced trial PMLE and MHE are identical, so that direct comparability of (2.33) and (2.34) are possible. A sparse balanced multicenter trial was simulated, with $n_i^T = n_i^C$ being generated from a Poisson distribution with Poisson parameter 5. The baseline parameter p_i^C was chosen from a uniform with parameters [0.1,0.3]. The risk ratio parameter was kept fixed for each simulation (replication size 10,000) and risk ratio values which ranged from 0.00001 to 3.3333 were considered. The number of centers k was chosen to be 20. The results for the two variance estimators are provided in Figure 2.7 for the bias of the variance estimators and in Figure 2.8 for the variance of the variance estimators indicating a slightly better behavior for (2.33) with respect to both criteria.

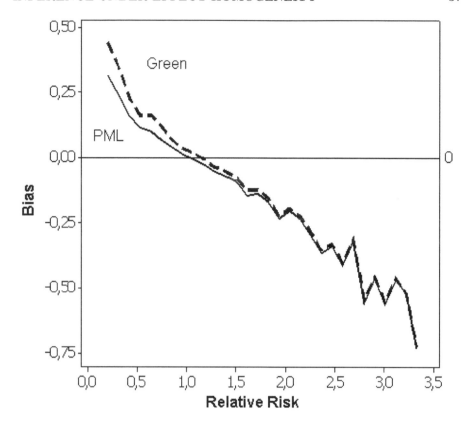

Figure 2.7 *A comparison of variance formulas provided by (2.33) (PML) and Greenland and Robins (2.34) (Green) for the sparse multicenter trial with respect to bias based upon a simulation*

Table 2.1 *Illustration of convergence of the fixed point procedure based upon the mapping $\Phi(\theta)$ for the data of table 1.5*

Iteration n	$\theta^{(n-1)}$	$\theta^{(n)}$	$\|\theta^{(n-1)} - \theta^{(n)}\|$
1	1.00000000	.88705370	.11294630
2	.88705370	.83522160	.05183208
3	.83522160	.81141940	.02380222
4	.81141940	.80048530	.01093411
5	.80048530	.79546170	.00502360
6	.79546170	.79315330	.00230843
7	.79315330	.79209270	.00106061
8	.79209270	.79160520	.00048751
9	.79160520	.79138110	.00022405
10	.79138110	.79127820	.00010288
11	.79127820	.79123100	.00004721
12	.79123100	.79120920	.00002182
13	.79120920	.79119930	.00000989

Table 2.2 *Illustration of convergence of the fixed point procedure based upon the mapping* $\Phi(\theta)$ *for the data of Table 1.11*

Iteration n	$\theta^{(n-1)}$	$\theta^{(n)}$	$|\theta^{(n-1)} - \theta^{(n)}|$
1	1.00000000	2.33891000	1.33891000
2	2.33891000	3.57786500	1.23895500
3	3.57786500	4.65437100	1.07650600
4	4.65437100	5.56571300	.91134170
5	5.56571300	6.32706400	.76135060
6	6.32706400	6.95822100	.63115790
7	6.95822100	7.47889400	.52067230
8	7.47889400	7.90700600	.42811250
9	7.90700600	8.25819800	.35119150
10	8.25819800	8.54580400	.28760620
11	8.54580400	8.78104100	.23523710
...	...	...	...
50	9.82232500	9.82240500	.00008011
51	9.82240500	9.82247000	.00006485
52	9.82247000	9.82252200	.00005245
53	9.82252200	9.82256500	.00004292
54	9.82256500	9.82259900	.00003433
55	9.82259900	9.82262700	.00002766
56	9.82262700	9.82265000	.00002289
57	9.82265000	9.82266800	.00001812
58	9.82266800	9.82268400	.00001621
59	9.82268400	9.82269800	.00001335
60	9.82269800	9.82270700	.00000954

Table 2.3 *Illustration of convergence of the fixed point procedure based upon the mapping* $\Gamma(\theta)$ *for the data of Table 1.11*

Iteration n	$\theta^{(n-1)}$	$\theta^{(n)}$	$|\theta^{(n-1)} - \theta^{(n)}|$
1	1.00000000	10.04283000	9.04283000
2	10.04283000	9.82654800	.21628280
3	9.82654800	9.82281700	.00373077
4	9.82281700	9.82275400	.00006294
5	9.82275400	9.82275200	.00000191

Table 2.4 *Number of steps needed for reaching the stop rule for the fixed mapping* $\Gamma(\theta)$ *and* $\Phi(\theta)$

Data	Using $\Gamma(\theta)$	Using $\Phi(\theta)$
Table 1.5	3	13
Table 1.8	4	15
Table 1.9	4	11
Table 1.10	3	13
Table 1.11	5	63
Table 1.12	2	2

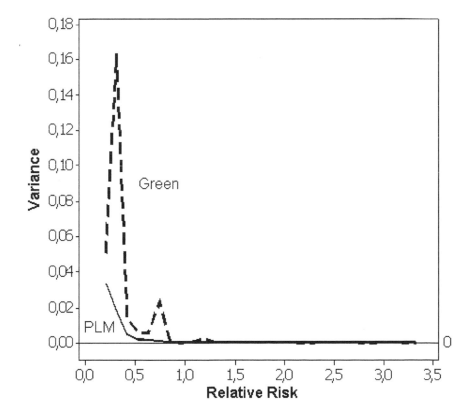

Figure 2.8 *A comparison of variance formulas provided by (2.33) (PML) and Greenland and Robins (2.34) (Green) for the sparse multicenter trial with respect to variance based upon a simulation*

CHAPTER 3

Modeling unobserved heterogeneity

In this chapter the focus is on coping with *unobserved heterogeneity*. This problem has received considerable attention in the literature. There appears to be a common understanding that unobserved heterogeneity is abundant in most meta-analyses. Engels et al. (2000) investigate 125 meta-analyses and conclude that

> heterogeneity was common regardless of whether treatment effects were measured by odds ratios or risk differences.

Approaches differ in the way they cope with this important problem. In the simplest approach, the variance of the pooled estimator is supplemented by an additional variance term, the *heterogeneity variance*. This heterogeneity variance can be estimated in various ways such as the moment approach (DerSimonian and Laird (1986), Malzahn et al. (2000), Böhning et al. (2002), Böhning et al. (2002) or Sidik and Jonkman (2005) or the maximum likelihood method (Hardy and Thompson (1996) or Whitehead and Whitehead (1991)). After the heterogeneity variance has been estimated the pooled estimator is recomputed using weights that incorporate the heterogeneity variance and a - usually enlarged - confidence interval is computed on basis of the incorporated heterogeneity variance. In addition, the random effects approach might be supplemented by a more complete modeling (see also Hardy and Thompson (1998) and Hedges and Vevea (1998)). A latent variable, an unobserved covariate might be supposed to be the source of this form of heterogeneity. Parametric approaches assume a parametric distribution for the latent variable (see Hardy and Thompson (1998), Martuzzi and Hills (1995)), and the approaches differ usually in the kind of parametric distribution that is assumed for the latent variable such as normal or Gamma. This latent variable might be a missing covariate (see Section 4) such as a treatment modification or different patient population, although other sources such as correlation of observations might be a source for this kind of heterogeneity. It is also discussed frequently whether the underlying risk is a source of heterogeneity for the relative risk (see Sharp et al. (1996), Sharp and Thompson (2000), Brensen et al. (1999), Egger and Smith (1995), van Houwelingen and Senn (1999), Thompson (1994), Arends et al. (2000)). It is pointed out here that the profile likelihood approach has the advantage of eliminating the effect of the baseline parameter before considering heterogeneity in the relative risk. In the following we develop a nonparametric random effects approach for modeling unobserved heterogeneity.

3.1 Unobserved covariate and the marginal profile likelihood

In this section, a general approach for coping with center-effect heterogeneity is proposed. Assume that the population of all centers consists of m subpopulations with weights q_j and subpopulation risk ratio θ_j. Let us consider again the likelihood (2.3) where we - for simplicity of presentation - consider only a single center:

$$Po(x^T, n^T p^C \theta) \times Po(x^C, n^C p^C) \qquad (3.1)$$

which becomes - after replacing p^C by their conditional maximum likelihood estimates $\frac{x^C + x^T}{n^C + \theta n^T}$

$$L(\mathbf{x}|\theta) = Po(x^T, n^T \frac{x^C + x^T}{n^C + \theta n^T} \theta) \times Po(x^C, n^C \frac{x^C + x^T}{n^C + \theta n^T}), \qquad (3.2)$$

where $Po(u, \lambda) = \exp(-\lambda)\lambda^u / u!$. Consider next the situation that for each observation $\mathbf{x} = (x^T, n^T, x^C, n^C)'$ there is an unobserved m-vector $\mathbf{y}_j$ with a 1 in the j-th position (and 0 otherwise) assigning the component population j to which the observation belongs to. Consider the joint density $f(\mathbf{x}, \mathbf{y}_j)$ of $\mathbf{x}$ and $\mathbf{y}_j$ which can be written as

$$f(\mathbf{x}, \mathbf{y}_j) = f(\mathbf{x}|\mathbf{y}_j) f(\mathbf{y}_j) = \prod_{j=1}^{m} (L(\mathbf{x}|\theta_j) q_j)^{y_j} \qquad (3.3)$$

where q_j is the probability, that observation $\mathbf{x}$ comes from component population $\mathbf{y}_j$ where the relative risk θ_j is valid. For details see McLachlan and Krishnan (1997) or McLachlan and Peel (2000). Note that (3.3) is an unobserved or latent likelihood. In these instances, the margin over the unobserved vector $\mathbf{y}_j$ is taken, leading to

$$\sum_{\mathbf{y}_j} \prod_{j=1}^{m} (L(\mathbf{x}|\theta_j) q_j)^{y_j} = \sum_{j=1}^{m} L(\mathbf{x}|\theta_j) q_j. \qquad (3.4)$$

In detail, the margin over the unobserved vector $\mathbf{y}_j$ leads to the *marginal density*

$$\sum_{j=1}^{m} Po(x^T, n^T \frac{x^C + x^T}{n^C + \theta n^T} \theta_j) \times Po(x^C, n^C \frac{x^C + x^T}{n^C + \theta_j n^T}) q_j \qquad (3.5)$$

where q_j is the weight, that the j-th population with parameter value θ_j receives. Taking now the log-likelihood over all centers and ignoring terms that do not involve the parameters the following *mixture profile log-likelihood* is achieved:

$$\sum_{i=1}^{k} \log \left[\sum_{j=1}^{m} \exp \left(-\frac{x_i n_i^T \theta_j}{n_i^C + \theta_j n_i^T} \right) \theta_j^{x_i^T} \times \exp \left(-\frac{x_i n_i^C}{n_i^C + \theta_j n_i^T} \right) \left(\frac{1}{n_i^C + \theta_j n_i^T} \right)^{x_i} q_j \right]$$

$$= \sum_{i=1}^{k} \log \left[\sum_{j=1}^{m} \exp(-x_i) \, \theta_j^{x_i^T} \left(\frac{1}{n_i^C + \theta_j n_i^T} \right)^{x_i} q_j \right]$$

$$\propto \sum_{i=1}^{k} \log \left[\sum_{j=1}^{m} \theta_j^{x_i^T} \left(\frac{1}{n_i^C + \theta_j n_i^T} \right)^{x_i} q_j \right] = L^*(Q)$$

$$(3.6)$$

which we may sometimes write as

$$L^*(Q) = \sum_{i=1}^{k} \log \left(\sum_{j=1}^{m} f_i(\theta_j) q_j \right) \tag{3.7}$$

where $f_i(\theta_j) = \frac{\theta_j^{x_i^T}}{(n_i^C + \theta_j n_i^T)^{x_i}}$ and $x_i = x_i^C + x_i^T$. Also, Q denotes the discrete probability distribution $Q = \begin{pmatrix} \theta_1 & \cdots & \theta_m \\ q_1 & \cdots & q_m \end{pmatrix}$ giving weights q_j to the risk ratio θ_j in subpopulation j, also called the *mixing distribution*.

3.2 Concavity, the gradient function and the PNMLE

It is easy to verify that $L^*(Q)$ is a *concave* function in the set Ω of all discrete probability distributions, though this is not necessarily the case for Ω_m, the set of all discrete probability distributions with *exactly* m support points (subpopulations). Hence, a global *profile mixture maximum likelihood estimator* (PNMLE) exists, but the number of support points is itself part of the estimation process. Let us propose the *gradient function* as an important tool for finding the PNMLE. In particular, set for arbitrary, but fixed $Q = \begin{pmatrix} \theta_1 & \cdots & \theta_m \\ q_1 & \cdots & q_m \end{pmatrix}$ and any $\theta > 0$, the gradient function is

$$d(\theta, Q) = \frac{1}{k} \sum_{i=1}^{k} \frac{f_i(\theta)}{\sum_{j=1}^{m} f_i(\theta_j) q_j} \tag{3.8}$$

We can define the gradient function (3.8) by means of the concept of the directional derivative where it is contained as the essential part (for details see Lindsay (1983), Lindsay (1995), Böhning (2000)). A first major application arises in the *general mixture maximum likelihood theorem* which states that $\hat{Q} = \begin{pmatrix} \hat{\theta}_1 & \cdots & \hat{\theta}_m \\ \hat{q}_1 & \cdots & \hat{q}_m \end{pmatrix}$ is PNMLE if and only if $d(\theta, \hat{Q}) \leq 1$ for all $\theta > 0$. In addition, for the support points of $\hat{Q}$ we have that the upper bound becomes sharp, e.g., $d(\hat{\theta}_j, \hat{Q}) = 1$. From this result, we can identify effect homogeneity without further testing. Indeed, let $\hat{\theta}_{PMLE}$ denote the profile maximum

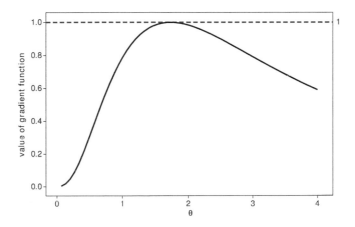

Figure 3.1 *Gradient function $d(\theta, \hat{\theta}_{PMLE})$ for Lidocaine trial (see also Table 1.7)*

likelihood estimator under homogeneity. If

$$d(\theta, \hat{\theta}_{PMLE}) = \frac{1}{k} \sum_{i=1}^{k} \frac{f_i(\theta)}{f_i(\hat{\theta}_{PMLE})} \leq 1,$$

for all $\theta > 0$, then $\hat{\theta}_{PMLE}$ must be the PNMLE, and no further search for heterogeneity is necessary.

Lidocaine trial. We come back to the multicenter study presented in Table 1.7. A graph of the gradient function $\theta \to d(\theta, \hat{\theta}_{PMLE})$ for the maximum likelihood estimator $\hat{\theta}_{PMLE}$ of θ under homogeneity (more precisely, the one-point probability measure giving all weight to $\hat{\theta}_{PMLE}$) is provided in Figure 3.1 showing clear evidence of homogeneity, making further testing for heterogeneity unnecessary.

3.2.1 Cholesterol lowering treatment and coronary heart disease

Let us consider again the multicenter study presented in Table 1.6. Here, the graph (see Figure 3.2) of the gradient function $\theta \to d(\theta, \hat{\theta}_{PMLE})$ for the maximum likelihood estimator $\hat{\theta}_{PMLE}$ of θ under homogeneity indicates clear

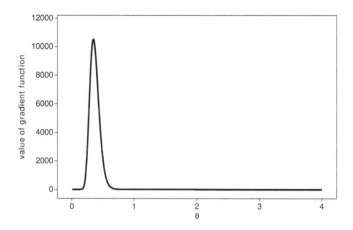

Figure 3.2 *Gradient function $d(\theta, \hat{\theta}_{PMLE})$ for cholesterol lowering trial (see also Table 1.6)*

evidence of heterogeneity. The upper bound one is violated (see Figure 3.2) and $\hat{\theta}_{PMLE}$ cannot be the PNMLE. In fact, it is clear that the PNMLE will have more than one support point.

For the general construction of the PNMLE, numerical algorithms are required.

3.3 The PNMLE via the EM algorithm

A major tool for constructing the maximum likelihood estimates is the EM algorithm (Dempster et al. (1977), McLachlan and Krishnan (1997)). It requires the specification of a suitable *complete data likelihood*, which for mixtures is conventionally taken as

$$\prod_{i=1}^{k}\prod_{j=1}^{m}\left(Po(x_i^T, n_i^T \frac{x_i^C + x_i^T}{n_i^C + \theta_j n_i^T}\theta_j) \times Po(x_i^C, n_i^C \frac{x_i^C + x_i^T}{n_i^C + \theta_j n_i^T}) \times q_j \right)^{y_{ij}}, \quad (3.9)$$

where $y_{ij} = 1$, if center i belongs to subpopulation j, and 0 otherwise. Since y_{ij} are unobserved, they are replaced in the *E-step* of the EM algorithm by their expected values

$$e_{ij} = E(Y_{ij}|Q, \text{ data})$$

$$= \frac{Po(x_i^T, n_i^T \frac{x_i^C + x_i^T}{n_i^C + \theta_j n_i^T} \theta_j) \times Po(x_i^C, n_i^C \frac{x_i^C + x_i^T}{n_i^C + \theta_j n_i^T}) \times q_j}{\sum_{j'=1}^{m} Po(x_i^T, n_i^T \frac{x_i^C + x_i^T}{n_i^C + \theta_{j'} n_i^T} \theta_{j'}) \times Po(x_i^C, n_i^C \frac{x_i^C + x_i^T}{n_i^C + \theta_{j'} n_i^T}) \times q_{j'}}. \tag{3.10}$$

Replacing y_{ij} in (3.9) by their expected values leads to the *expected* complete data likelihood

$$\prod_{i=1}^{k} \prod_{j=1}^{m} \left(Po(x_i^T, n_i^T \frac{x_i^C + x_i^T}{n_i^C + \theta_j n_i^T} \theta_j) \times Po(x_i^C, n_i^C \frac{x_i^C + x_i^T}{n_i^C + \theta_j n_i^T}) \times q_j \right)^{e_{ij}}, \tag{3.11}$$

which can be maximized in θ_j and q_j, separately. This established the *M-step* of the EM algorithm. In fact, we find easily that

$$\hat{q}_j = \frac{1}{k} \sum_{i=1}^{k} e_{ij}.$$

Furthermore, $\hat{\theta}_j$ can be found from the equation

$$\sum_{i=1}^{k} \frac{e_{ij} x_i^T}{\theta_j} - \frac{e_{ij} x_i n_i^T}{n_i^C + \theta_j n_i^T} = 0$$

by using the iteration

$$\hat{\theta}_j = \frac{\sum_{i=1}^{k} e_{ij} x_i^T n_i^C w_i(\hat{\theta}_j)}{\sum_{i=1}^{k} e_{ij} x_i^C n_i^T w_i(\hat{\theta}_j)} \tag{3.12}$$

with $w_i(\theta) = 1/(n_i^T \theta + n_i^C)$, in analogy to the homogenous case (2.14).

3.4 The EMGFU for the profile likelihood mixture

When the gradient function indicates heterogeneity, usually the number of components adequate to model this heterogeneity will be unknown and several values for m need to be considered. Hence, it appears appropriate to consider *all* possible values of m, starting from $m = 1$ to the number of components involved in the PNMLE. The following algorithm is similar to the *EM algorithm with a gradient function update* (EMGFU) (Böhning (2003)).

The initial step starts with the case of homogeneity ($m = 1$) and the computation of the profile maximum likelihood estimator under homogeneity. If the gradient function violates the upper bound, e.g., $d(\theta_{max}, \theta_{PMLE}) > 1$, then the number of components is increased to $m = 2$ and the EM algorithm of the previous section is utilized with initial values for the two components $\theta_1 = \theta_{PMLE}$ and $\theta_2 = \theta_{max}$ to compute a discrete two-support point probability distribution $Q^{(2)}$. Otherwise (if the gradient function does not violate the upper bound), the algorithm is stopped.

Now, suppose that the EM algorithm has generated a current value of m for a discrete probability distribution $Q^{(m)}$ which has support points $\theta_1, ..., \theta_m$. If the gradient function violates the upper bound, e.g., $d(\theta_{max}, Q^{(m)}) > 1$, then the number of components is increased to $m = m + 1$ and the EM algorithm of the previous section is utilized with initial values for the $m + 1$ components $\theta_1, ..., \theta_m$ and $\theta_{m+1} = \theta_{max}$ to compute a discrete $(m+1)$-support point probability distribution $Q^{(m+1)}$. Otherwise (if the gradient function does not violate the upper bound), the algorithm is stopped. This step is repeated until the PNMLE is reached. The advantage of the EMGFU lies in the fact that it combines a strategy for generating the nonparametric profile maximum likelihood estimator with a search for the best local mixture maximum likelihood estimator with exactly m components.

3.4.1 Cholesterol lowering treatment and coronary heart disease.

We would like to demonstrate the EMGFU for the multicenter trial given in Table 1.6). In this case, we had found clear evidence of heterogeneity (see Section 3.2). Table 3.1 provides details on this analysis. The EMGFU algorithm starts with homogeneity and provides $\theta_{PMLE} = 0.9716$, then increases the number of support points stepwise by means of the gradient function until the nonparametric profile maximum likelihood estimator with $m = 4$ components is reached. Indeed, from the gradient function plot (Figure 3.3), it is evident, that the global nonparametric profile maximum likelihood estimator has been reached.

3.5 Likelihood ratio testing and model evaluation

Profile likelihoods behave similarly to likelihoods. However, for mixture models this just means that we have to face the same problems. Profile likelihood ratios will not have standard χ^2 distributions, so that choices for the number of components, solely based upon the likelihood ratio, might be misleading. Then should be accompanied by other selection criteria such as the *Bayesian Information Criterion* (BIC) which has proved to be a valuable selection criterion in other settings. McLachlan and Peel (2000) discuss and compare various selection criteria. Within the simpler criteria, the *Akaike Information Criteria* (AIC) shows a tendency to select too many components (overestimate m), whereas the BIC, though not always correct, behaves better. In Table 3.1, it can be seen that the best model according to BIC is the model with two components.

Table 3.1 Results of the mixture model fitting for the trials of cholesterol lowering treatment and coronary heart disease given in Table 1.6; $\hat{Q}^{(m)}$ is the mixture
maximum profile likelihood estimate with m components

m	$\hat{\theta}_j$	$\hat{q}_j$	$d(\theta_{max}, \hat{Q}^{(m)})$	$L^*(\hat{Q}^{(m)})$	BIC
1	0.9716	1	10,518.11	$-50,172.01$	$-100,347.52$
2	1.0058	0.8901	1.7856	$-50,161.54$	$-100,333.57$
	0.4401	0.1099			
3	0.9776	0.8029	1.2998	$-50,160.62$	$-100,338.72$
	0.4283	0.1013			
	1.2827	0.0958			
4	1.0016	0.6546	1.0000	$-50,159.66$	$-100,343.79$
PNMLE	0.3665	0.0558			
	0.6962	0.1916			
	1.2793	0.0980			

3.6 Classification of centers

Note that for i fixed, e_{ij}, as given in (3.10), is a probability distribution. In
fact, e_{ij} is the posterior probability for center i belonging to subpopulation
j. This enables classification of center i into that subpopulation j where the
posterior probability is the largest. In the cholesterol lowering trial there was
a PNMLE found consisting of four subpopulations (see Table 3.1. However,
according to the BIC, only two subpopulations are required. Table 3.2 provides
the posterior probabilities for each of the 33 centers including a classification
of each center into the component, associated with the highest posterior.

3.7 A reanalysis on the effect of beta-blocker after myocardial infarction

Returning to the MAIPD for studying the effect of beta-blockers treatment
for reducing mortality after myocardial infarction (Yusuf et al. (1985)) is presented in Table 1.5 in Chapter 1. A few important aspects were already mentioned in Chapter 1:

- It can be easily seen which of the studies have positive and which have
  negative effects.

- Significant positive effects can be easily identified by finding those studies

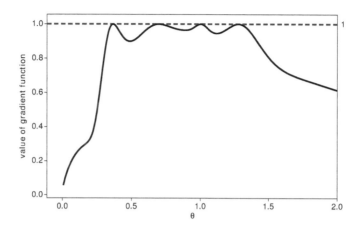

Figure 3.3 *Gradient function $d(\theta, \hat{Q}_{PNMLE})$ (see also Table 3.1) for cholesterol lowering trial (see also Table 1.6)*

with confidence intervals above the no-effect line, the horizontal line at zero. Here, there are none.

- Significant negative effects can also be easily identified by finding those studies with confidence intervals below the no-effect line. These are the four studies: 7, 10, 21, and 22.

- Some studies show strong negative effects (stronger than the studies 7 and 10 which had a significant negative effect) but are *not* significant, since there is a large variance associated with these studies due to small sizes. This refers to studies 2, 3, 6, and 13.

These points cause three questions of interest to arise:

- a) Is there an effect of treatment?

- b) Is this effect homogeneous? In particular, does study 14 with its borderline significant harmful effect provide enough empirical evidence to create a positive effect cluster, potentially including also other studies such as numbers 5, 17, 18, and 19?

  - b1) If it is homogeneous, how large is its effect size?

— b2) If the effect is heterogeneous, what are the clusters involved?

An analysis of unobserved heterogeneity shows a very mild form of heterogeneity: a mixture model with two components provides the nonparametric maximum likelihood estimate. The results are provided in Table 3.3. If we compare the log-likelihoods for the homogeneity model and the two-component mixture model we find a value for the difference of 0.4 and a considerably higher BIC value for the homogeneity model. We can conclude that there is a clear and significant beneficial effect of beta-blockers therapy, at least on the basis of those data. See also Figure 3.4 for an illustration.

Although there appears to be no significant second cluster, it might be interesting to see which studies would have been allocated to the second cluster using the MAP rule. We find only study 14 would have been allocated to the second component. This underlines the clarity of the resulting analysis.

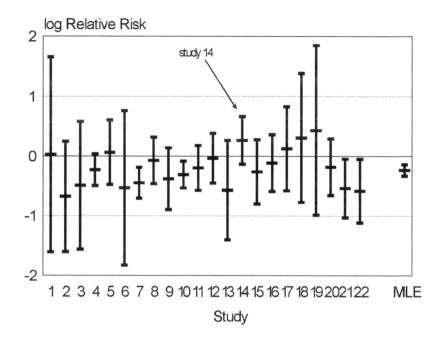

Figure 3.4 *Effect of beta-blockers for reducing mortality after myocardial infarction (Yusuf et al. (1985)) in 22 studies expressed as log-relative risk with 95% confidence interval and MLE as summary measure with 95% confidence interval*

Table 3.2 Classification of centers for the cholesterol lowering treatment trials given in Table 1.6 according to the posterior distribution when using the two-component mixture $\hat{Q}^{(2)}$ (see Table 3.1)

| | posterior of component | | classified as |
Center	1	2	belonging to
1	0.09775	0.902248	2
2	0.99993	0.000069	1
3	0.8994	0.100599	1
4	0.73118	0.268823	1
5	0.78973	0.21027	1
6	0.97666	0.023345	1
7	0.00002	0.999979	2
8	0.98652	0.01348	1
9	1	0	1
10	0.99993	0.000068	1
11	0.63296	0.367043	1
12	0.99889	0.001115	1
13	0.99951	0.000493	1
14	0.82041	0.179595	1
15	1	0	1
16	1	0	1
17	0.99498	0.005023	1
18	0.99958	0.000422	1
19	0.91806	0.081936	1
20	0.98398	0.016024	1
21	1	0.000001	1
22	0.98862	0.011379	1
23	0.90455	0.09545	1
24	1	0	1
25	0.99441	0.005587	1
26	0.88153	0.118473	1
27	0.99956	0.000443	1
28	1	0.000005	1
29	0.99997	0.000029	1
30	0.99963	0.000369	1
31	1	0	1
32	0.88138	0.118618	1
33	0.89578	0.104223	1

Table 3.3 Results of the mixture model fitted for the MAIPD on the effect of beta-blockers treatment for reducing mortality after myocardial infarction (Yusuf et al. (1985)). $\hat{Q}^{(m)}$ is the mixture maximum profile likelihood estimate with m components

m	$\hat{\theta}_j$	$\hat{q}_j$	$L^*(\hat{Q}^{(m)})$	BIC
1	0.7912	1	$-12,979.3$	$-25,961.7$
2	0.7685	0.9054	$-12,978.9$	$-25,967.2$
PNMLE	1.1290	0.0946		

CHAPTER 4

Modeling covariate information

The modeling of unobserved heterogeneity may be useful when heterogeneity of treatment effect is present, but it cannot obviously unmask the possible reasons for heterogeneity between study results. In particular, understanding the reasons for some heterogeneity is more important than the evidence for its existence.

This chapter provides a review of classical methods for investigating the reasons for heterogeneity of the treatment effects and points out some potential problems related to them. Additionally, this chapter presents a novel model for incorporating covariate information which is modeling of covariate information using the profile likelihood approach and the applications of this model.

4.1 Classical methods

Recent methods have been suggested to incorporate modeling of covariate information to investigate the reasons for heterogeneity and to estimate the treatment effect based upon the significant covariates. More recently, there has been discussion on the choice of appropriate statistical methods to address this issue. This section provides a review of classical methods and points out some potential problems related to them at the end of the section.

4.1.1 Weighted regression

These methods use the idea of a generalized linear model to explain the variation of treatment effects by covariate information (for example, see Hedges (1994b), Raudenbush (1994), Berkey et al. (1995), Stram (1996), Thompson and Sharp (1999)). Let y_i be the observed treatment effect in the i-th study, for $i = 1, \ldots, k$. The y_i can be the observed log-odds ratio or log-relative risk in a trial with binary outcome or the observed mean difference in a trial with continuous outcome. In our case, y_i is the observed log-relative risk. It is assumed that y_i are independently distributed as

$$y_i \sim N(\theta_i, v_i) \tag{4.1}$$

where θ_i is the true log-relative risk in the i-th study and v_i is the variance of the log-relative risk in the i-th study.

For the fixed-effects model, it is supposed that there are p known covariates $z_1, \ldots, z_p$ which are presumed to account completely for variation in the true log-relative risk, so θ_i is specified by $\beta' \mathbf{z}_i$, where β is a column vector of regression coefficients $(\beta_0, \beta_1, \ldots, \beta_p)'$ and $\mathbf{z}_i$ is a column vector that contains the values of p covariates for study i.

$$y_i \sim N(\beta' \mathbf{z}_i, v_i) \tag{4.2}$$

and the fixed-effects regression model for log-relative risk estimate becomes:

$$y_i = \beta' \mathbf{z}_i + e_i \tag{4.3}$$

where e_i is the error of estimation of study i. Each e_i is statistically independent with a mean of zero and variance of v_i.

For the random-effects model, it is assumed that the true log-relative risk θ_i vary randomly across studies and are independently distributed as:

$$\theta_i \sim N(\mu, \tau^2) \tag{4.4}$$

where μ is the mean of the distribution of θ_i across studies, and τ^2 is the variance of the distribution of θ_i across studies. To incorporate the covariates and thus account for heterogeneity among studies, μ can be specified by $\beta' \mathbf{z}_i$

$$\theta_i \sim N(\beta' \mathbf{z}_i, \tau^2). \tag{4.5}$$

The random-effects regression model for log-relative risk estimate becomes:

$$y_i = \beta' \mathbf{z}_i + \delta_i + e_i \tag{4.6}$$

where δ_i are the random effects of study i, that is the deviation of study i's true treatment effect from the true mean of all studies having the same covariate values. Each random effect, δ_i, is assumed to be independent with a mean of zero and variance τ^2, and e_i is the error of estimation of study i. Each e_i is statistically independent with a mean of zero and variance of v_i. The design vector $\mathbf{z}_i$ and within-study variance v_i are assumed to be known. Additionally, regression coefficient vector β and between-study variance τ^2 are estimated from the data. Notice that equation (4.6) has two components in its error term, $\delta_i + e_i$, which are assumed to be independent, leading to a covariance equals zero, so that the marginal variance of y_i is

$$v_i^* = Var(\delta_i + e_i) = \tau^2 + v_i. \tag{4.7}$$

The statistical literature describes equation (4.6) as a mixed effects linear model with fixed effects β and random effects δ_i.

A particular disadvantage of this modeling is the inappropriate identity link which is used to link covariate information to relative risk, since it does not guarantee that the relative risk estimates are positive which would be an essential requirement for a relative risk. This problem is overcome by using the canonical link which guarantees that relative risk estimates are positive. The second disadvantage is a potential violation of the normality assumption of both the observed treatment effects and the random effects. For example, it has been assumed that the log-relative risk is normally distributed, however, this may not be appropriate for small studies or small number of events. Moreover, in practice, the v_i are rather estimated from the data than known, so the correlation between estimates of log-relative risk and their variance estimates may produce bias in the estimates of regression coefficients (Berkey et al. (1995)). These problems are overcome by directly using the structure of the binary data, binomial or Poisson.

4.1.2 Logistic regression

Another classical method uses directly the binomial structure for the binary data (for example, see Aitkin (1999b), Thompson and Sharp (1999), Turner et al. (2000)). Let y_{ij} be the number of events in the j-th group ($j = 0$ control, $j = 1$ treated) of study i and n_{ij} be the number of subjects in the j-th group of study i. Also let π_{ij} denote the risk (probability) of an event in the j-th group of study i. It is assumed that y_{ij} is independently distributed as:

$$y_{ij} \sim \text{Binomial}(\pi_{ij}, n_{ij}). \tag{4.8}$$

Suppose there are p known covariates $z_1, \ldots, z_p$ which might be the sources of variation between studies. Let $\mathbf{z}_{ij}$ be a column vector that contains the values of p covariates in the j-th group of study i, and u_j be an indicator variable for the treatment group (0 for control, 1 for treated). The conventional logistic regression model can be written as:

$$\text{logit}(\pi_{ij}) = \alpha_i + \beta^* u_j + \beta' \mathbf{z}_{ij} \tag{4.9}$$

where α_i is the intercept parameter in study i, β^* is the overall average value of log-odds ratio adjusted for covariates, and β is a column vector that contains the log-odds ratio per unit change for p covariates. For the conventional logistic regression model, it is assumed that α_i is a fixed parameter and β^* is a fixed effects parameter. However, there is no allowance for heterogeneity effect in this model.

One method to incorporate heterogeneity effect into the models is the multi-level approach. An appropriate model can be written as

$$\text{logit}(\pi_{ij}) = \alpha_i + \beta_i^* u_j + \beta' \mathbf{z}_{ij} \tag{4.10}$$

where α_i is the fixed intercept parameter in study i, and β_i^* is the log-odds ratio in study i which varies randomly across studies and has an independent normal distribution as:

$$\beta_i^* \sim N(\beta^*, \tau^2) \tag{4.11}$$

Notice that in equations (4.9) and (4.10), it is assumed that α_i is a fixed parameter. However, the number of α_i parameters increases with the number of centers, leading to the Neyman-Scott problem (see Neyman and Scott (1948)).

An alternative multilevel model in which intercept parameters α_i are regarded as random rather than fixed is written as follows:

$$\text{logit}(\pi_{ij}) = \alpha_i + \beta_i^* u_j + \beta' \mathbf{z}_{ij} \tag{4.12}$$

where the α_i are independently distributed as:

$$\alpha_i \sim N(\alpha, \tau_\alpha^2) \tag{4.13}$$

and where the β_i^* are independently distributed as:

$$\beta_i^* \sim N(\beta^*, \tau_\beta^2). \tag{4.14}$$

and also $cov(\alpha_i, \beta_i^*) = \rho \tau_\alpha \tau_\beta$, where ρ is the correlation coefficient. An alternative formula of equation (4.12) can be written as:

$$\text{logit}(\pi_{ij}) = \alpha + \gamma_i + \beta^* u_j + \delta_i u_j + \beta' \mathbf{z}_{ij} \tag{4.15}$$

where the γ_i are independently distributed as

$$\gamma_i \sim N(0, \tau_\alpha^2) \tag{4.16}$$

and where the δ_i are independently distributed as

$$\delta_i \sim N(0, \tau_\beta^2). \tag{4.17}$$

This is the simplest and most conventional multilevel model together with the random-effects model for β_i^* which could lead to an extension of the univariate random effects model to a bivariate normal model. It is important to consider the covariance between the α_i and β_i^* in a bivariate normal model. If $cov(\alpha_i, \beta_i^*)$ is assumed to be zero, the between-study variance of the log-odds across control groups is equal to τ_α^2. Meanwhile, that across treatment groups is equal to $\tau_\beta^2 + \tau_\alpha^2$. The between-study variation in control groups is thereby forced to be less than or equal to the between-study variation in treatment groups. This assumption may not be appropriate for the general situation. When $cov(\alpha_i, \beta_i^*)$ is estimated rather than assumed to be zero, the variance-covariance matrix of the bivariate log-odds parameter estimates is modeled by a combination of the three parameters τ_α, τ_β, and ρ. This nonzero covariance assumption allows the model to investigate a relationship between baseline risk and treatment effect. However, this alternative model presents an extended complexity in the multilevel approach. The above issue has been discussed in Turner et al. (2000).

Notice that in equations (4.9), (4.10), and (4.12), two kinds of parameters occur. The first type is the parameter of interest, that is the coefficient of indicator variable β^* and the coefficients of covariates β. The second type is the nuisance parameter, that is the intercept parameter α_i. However, the nuisance parameter is not our main parameter of interest, but it is a parameter that gives a complete description of the model while complicating it. All these models, based on the binomial structure of data, suffer from dealing with the nuisance parameter α_i.

4.2 Profile likelihood method

For the reasons mentioned above, the profile likelihood method which is a traditional method of dealing with nuisance parameters, (see, for example, Aitkin (1998), Murphy and Van der Vaart (2000), or Pawitan (2001)) becomes attractive for modeling of covariate information. A novel model for modeling of covariate information based upon a modification of the generalized linear model using the profile likelihood method is presented in this section.

4.2.1 A generalized linear model

The theory of generalized linear model was first developed by Nelder and Wedderburn (1972). In our case, a generalized linear model (McCullagh and Nelder (1989)) has been applied to explain the variation of treatment effect by means of covariate information. Let z_{ij} be the value of the j-th covariate in the i-th center, for $i = 1, \ldots, k$ and $j = 1, \ldots, p$. The linear predictor η_i for covariate information in the i-th center can be defined as:

$$\eta_i = \beta_0 + \beta_1 z_{i1} + \beta_2 z_{i2} + \ldots + \beta_p z_{ip} \tag{4.18}$$

where $\beta_0, \beta_1, \ldots, \beta_p$ are the parameters of the model. We consider again the profile log-likelihood for the relative risk estimator:

$$L^*(\theta) = \sum_{i=1}^{k} x_i^T \log(\theta_i) - (x_i^C + x_i^T) \log(n_i^C + \theta_i n_i^T). \tag{4.19}$$

Here we use the canonical link $\theta_i = \exp(\eta_i)$ to link the linear predictor (4.18) to the relative risk parameter θ_i in (4.19) which guarantees that $\theta_i \geq 0$ which is an essential requirement for a relative risk. With the canonical link, the profile log-likelihood for covariate information becomes

$$L^*(\beta) = \sum_{i=1}^{k} x_i^T \eta_i - (x_i^C + x_i^T) \log(n_i^C + \exp(\eta_i) n_i^T) \tag{4.20}$$

with $\eta_i = \beta_0 + \beta_1 z_{i1} + \beta_2 z_{i2} + \ldots + \beta_p z_{ip}$.

4.2.2 Finding maximum likelihood estimates

We now estimate the parameters of model β_j, for $j = 0, 1, \ldots, p$. To find the maximum likelihood estimate of β_j we need to maximize the profile log-likelihood (4.20). For this purpose, consider the partial derivative w.r.t β_j

$$\frac{\partial L^*}{\partial \beta_j}(\beta) = \sum_{i=1}^{k} x_i^T z_{ij} - x_i n_i^T \frac{\exp(\eta_i)}{n_i^C + \exp(\eta_i) n_i^T} z_{ij} \tag{4.21}$$

where $x_i = x_i^T + x_i^C$, and the corresponding vector of partial derivatives, the gradient is:

$$\nabla L^*(\beta) = (\frac{\partial L^*}{\partial \beta_0}, \cdots, \frac{\partial L^*}{\partial \beta_p})'. \tag{4.22}$$

Furthermore, *Hesse* matrix of second derivatives is:

$$\frac{\partial^2 L^*}{\partial \beta_h \partial \beta_j}(\beta) = -\sum_{i=1}^{k} \frac{x_i n_i^T n_i^C \exp(\eta_i)}{(n_i^C + \exp(\eta_i) n_i^T)^2} z_{ij} z_{ih} \tag{4.23}$$

so that (4.23) becomes in matrix form:

$$\nabla^2 L^*(\beta) = \left(\frac{\partial^2 L^*}{\partial \beta_h \partial \beta_j}(\beta) \right) = -Z'W(\beta)Z \tag{4.24}$$

where Z is the design matrix of covariate information defined as:

$$Z = \begin{pmatrix} z_{10} & z_{11} & z_{12} & \cdots & z_{1p} \\ z_{20} & z_{21} & z_{22} & \cdots & z_{2p} \\ . & . & . & \cdots & . \\ z_{k0} & z_{k1} & z_{k2} & \cdots & z_{kp} \end{pmatrix}_{k \times (p\,|\,1)}$$

with

k	is the number of centers,
p	is the number of covariates,
z_{i0}	is the constant value of coefficient in the i-th center, $z_{i0} = 1$,
$z_{i1}, \ldots, z_{ip}$	is value of covariates in the i-th center,

and $W(\beta)$ is a diagonal matrix defined as:

$$W(\beta) = \begin{pmatrix} w_{11} & w_{12} & w_{13} & \cdots & w_{1k} \\ w_{21} & w_{22} & w_{23} & \cdots & w_{2k} \\ . & . & . & \cdots & . \\ w_{k1} & w_{k2} & w_{k3} & \cdots & w_{kk} \end{pmatrix}_{k \times k}$$

with $w_{ij} = 0$, if $i \neq j$ and

$$w_{ii} = \frac{x_i n_i^T n_i^C \exp(\eta_i)}{(n_i^C + \exp(\eta_i) n_i^T)^2}.$$

Then, we used the Newton-Raphson procedure to iteratively construct the maximum likelihood estimates of β_j. Choose some $\beta^{(0)}$ as initial values (for example $\beta^{(0)} = 0$) and then update β according to:

$$\beta^{(n+1)} = \beta^{(n)} - \nabla^2 L^*(\beta^{(n)})^{-1} \nabla L^*(\beta^{(n)}) \tag{4.25}$$

until convergence.

4.2.3 Finding standard errors of effect estimates

We can estimate variances of maximum likelihood estimates from the negative inverse of the information matrix (4.24). The variances estimate equals:

$$v\hat{a}r(\hat{\beta}_j) = (Z'W(\hat{\beta})Z)^{-1}_{jj} \qquad (4.26)$$

so that the standard errors become:

$$s\hat{.}e.(\hat{\beta}_j) = \sqrt{v\hat{a}r(\hat{\beta}_j)} \qquad (4.27)$$

and significance of individual effects can be consequently obtained by means of a Wald-test defined as:

$$T_j = \frac{\hat{\beta}_j}{s\hat{.}e.(\hat{\beta}_j)} \qquad (4.28)$$

so that the P-value of Wald-test under the null hypothesis of no effect of the j-th covariate, can be found as:

$$P\text{-value} = 1 - F(T_j) \quad , T_j \geq 0$$

with $F(T_j)$ as the cumulative function of the standard normal distribution.

4.2.4 Finding relative risk and 95% CI for covariate information

After finding standard errors of effect estimates, we compute the relative risk and 95% CI for covariate information as follows:

We consider again the linear predictor model (4.18):

$$\eta_i = \beta_0 + \beta_1 z_{i1} + \ldots + \beta_p z_{ip}. \qquad (4.29)$$

If $z_{i1}, z_{i2}, \ldots, z_{ip} = 0$, then $\eta_i = \beta_0$, and the relative risk for the i-th center is estimated as:

$$\hat{\theta}_i = \exp(\hat{\beta}_0). \qquad (4.30)$$

The associated 95% CI can be defined as:

$$\exp\{\hat{\beta}_0 \pm 1.96 \times s\hat{.}e.(\hat{\beta}_0)\}. \qquad (4.31)$$

If $z_{i1}, z_{i2}, \ldots, z_{ip} \neq 0$, then $\eta_i = \beta_0 + \beta_1 z_{i1} + \ldots + \beta_p z_{ip}$, and the relative risk for the i-th center is estimated as:

$$\hat{\theta}_i = \exp(\hat{\eta}_i) \qquad (4.32)$$

where $\hat{\eta}_i = \hat{\beta}_0 + \hat{\beta}_1 z_{i1} + \ldots + \hat{\beta}_p z_{ip}$. The associated 95% CI can be defined as:

$$\exp\{\hat{\eta}_i \pm 1.96 \times s\hat{.}e.(\hat{\eta}_i)\} \qquad (4.33)$$

where $s\hat{.}e.(\hat{\eta}_i)$ can be defined as

$$s\hat{.}e.(\hat{\eta}_i) = \sqrt{z_i' \widehat{COV}(\hat{\beta}) z_i} \qquad (4.34)$$

where $z_i = (z_{i0}, z_{i1}, z_{i2}, \ldots, z_{ip})'$, with $z_{i0} = 1$, and $\widehat{COV}(\hat{\beta}) = (Z'W(\hat{\beta})Z)^{-1}$.

4.3 Applications of the model

Frequently, a MAIPD or a multicenter trial does not only provide information on treatment and control arms, outcome and sample size. However, it also includes potentially important coinformation. This might show some joint variation with the treatment effect. This section illustrates the applications of the modeling of covariate information using the profile likelihood approach to two examples of meta-analysis of clinical trials and one sample of a multicenter trial.

4.3.1 Quitting smoking

The first example is the meta-analysis of 59 trials that evaluates the effect of nicotine replacement therapy known as NRT on quitting smoking. These data are taken from DuMouchel and Normand (2000). It is of interest to find out whether NRT helps a person quit smoking. However, there are two different forms of NRT (patch and gum) and two different types of support (high support and low support) in the quitting smoking study. The data of quitting smoking study in 59 trials are displayed in Table 4.1. It has been determined that *NRT* is the binary covariate to describe the form of NRT; patch (NRT=1) or gum (NRT=0); and *Support* is the binary covariate to describe the type of support; high (Support=1) or low (Support=0).

The results of fitting the various models to the quitting smoking data are presented in Table 4.2. A forward selection procedure and profile likelihood ratio test (PLRT) have been applied to select the significant covariates. The critical value (5% level) of PLRT with 1 df, which is 3.841, has been used for selecting among possible models. The results indicate that the form of NRT yields the only significant change for the treatment effect of quitting smoking. The estimation of relative risk and 95% CI are presented in Table 4.3. According to the model without covariates, the relative risk equals 1.57, it means that the treatment can increase the success of quitting smoking by an average of 57%. Referring to the model with covariate the form of NRT, the relative risk equals 1.85 if a patch treatment is used and the relative risk equals 1.47 if a gum treatment is used. This suggests that there is an increase of quitting smoking of 85% in patch groups and an increase of quitting smoking of 47% in gum groups. This shows that simply combining the results from a meta-analysis or multicenter trial into an overall estimate is misleading.

Table 4.1 *Count data and characteristics of 59 trials on the efficacy of nicotine replacement therapy on quitting smoking*

Study	Name	Year	x^T	n^T	x^C	n^C	NRT	Support
1	Puska	1979	29	116	21	113	0	1
2	Malcom	1980	6	73	3	121	0	1
3	Fagerstrom	1982	30	50	23	50	0	1
4	Fee	1982	23	180	15	172	0	1
5	Jarvis	1982	22	58	9	58	0	1
6	Hjalmarson	1984	31	106	16	100	0	1
7	Killen	1984	16	44	6	20	0	1
8	Schneider	1985	9	30	6	30	0	1
9	Hall	1987	30	71	14	68	0	1
10	Tonnesen	1988	23	60	12	53	0	1
11	Blondal	1989	37	92	24	90	0	1
12	Garcia	1989	21	68	5	38	0	1
13	Killen	1990	129	600	112	617	0	1
14	Nakamura	1990	13	30	5	30	0	1
15	Campbell	1991	21	107	21	105	0	1
16	Jensen	1991	90	211	28	82	0	1
17	McGovern	1992	51	146	40	127	0	1
18	Pirie	1992	75	206	50	211	0	1
19	Zelman	1992	23	58	18	58	0	1
20	Herrera-1	1995	37	76	17	78	0	1
21	Buchkremer	1981	11	42	16	89	1	1
22	Hurt	1990	8	31	6	31	1	1
23	Ehrsam	1991	7	56	2	56	1	1
24	Tnsg	1991	111	537	31	271	1	1
25	Sachs	1993	28	113	10	107	1	1
26	Westman	1993	16	78	2	80	1	1
27	Fiore-1	1994	15	44	9	43	1	1
28	Fiore-2	1994	10	57	4	55	1	1
29	Hurt	1994	33	120	17	120	1	1

Note. x^T=number of smokers who quit smoking in the treatment group,
n^T=number of smokers in the treatment group,
x^C=number of smokers who quit smoking in the control group,
n^C=number of smokers in the control group.

Table 4.1 *Count data and characteristics of 59 trials on the efficacy of nicotine replacement therapy on quitting smoking*

Study	Name	Year	x^T	n^T	x^C	n^C	NRT	Support
30	ICRF	1994	76	842	53	844	1	1
31	Richmond	1994	40	160	19	157	1	1
32	Kornitzer	1995	19	150	10	75	1	1
33	Stapleton	1995	77	800	19	400	1	1
34	Campbell	1996	24	115	17	119	1	1
35	BR SOCIETY	1983	39	410	111	1208	0	0
36	Russell	1983	81	729	78	1377	0	0
37	Fagerstrom	1984	28	106	5	49	0	0
38	Jamrozik	1984	10	101	8	99	0	0
39	Jarvik	1984	7	25	4	23	0	0
40	Clavel-Chapel	1985	24	205	6	222	0	0
41	Schneidera	1985	2	13	2	23	0	0
42	Page	1986	9	93	13	182	0	0
43	Campbell	1987	13	424	9	412	0	0
44	Sutton	1987	21	270	1	64	0	0
45	Areechon	1988	56	99	37	101	0	0
46	Harackiewicz	1988	12	99	7	52	0	0
47	Llivina	1988	61	113	28	103	0	0
48	Sutton	1988	5	79	2	82	0	0
49	Gilbert	1989	11	112	9	111	0	0
50	Hughes	1989	23	210	6	105	0	0
51	Hughes	1990	15	59	5	19	0	0
52	Mori	1992	30	178	22	186	0	0
53	Nebot	1992	5	106	13	319	0	0
54	Fortmann	1995	44	262	42	261	0	0
55	Abelin	1989	17	100	11	99	1	0
56	Daughton	1991	28	106	4	52	1	0
57	Tonneson	1991	17	145	2	144	1	0
58	Burton	1992	29	115	22	119	1	0
59	Paoletti	1996	15	60	4	60	1	0

Note. x^T=number of smokers who quit smoking in the treatment group,
n^T=number of smokers in the treatment group,
x^C=number of smokers who quit smoking in the control group,
n^C=number of smokers in the control group.

Table 4.2 *Results of fitting various models to the meta-analysis of quitting smoking study*

$L^*(\hat{\beta})$	Covariates	$\hat{\beta}_j$	S.E.	P-value	LR-Test
-17218.8100	Intercept	0.4483	0.0392	0.0000	
-17215.4000^d	Intercept	0.3850	0.0459	0.0000	6.8200^a
	NRT	0.2301	0.0887	0.0047	
-17218.7300	Intercept	0.4700	0.0647	0.0000	0.1600^a
	Support	-0.0343	0.0813	0.3367	
-17214.8400	Intercept	0.4356	0.0661	0.0000	1.1200^b
	NRT	0.2526	0.0912	0.0028	
	Support	-0.0893	0.0838	0.1434	
-17214.6500	Intercept	0.4222	0.0697	0.0000	0.3800^c
	NRT	0.3558	0.1950	0.0340	
	Support	-0.0657	0.0926	0.2391	
	NRT*Support	-0.1328	0.2207	0.2738	

[a] Comparison of the current model with the model without covariates.
[b] Comparison of the current model with the model with the covariate NRT.
[c] Comparison of the current model with the model with both covariates NRT and support.
[d] Selected model for the meta-analysis of quitting smoking study.

Table 4.3 *Results of estimation of RR and 95% CI of the meta-analysis of quitting smoking study*

	S.E.	RR (95% CI)
No covariates	0.0392	1.5656 (1.4500, 1.6905)
Form of NRT		
Patch	0.0759	1.8499 (1.5942, 2.1466)
Gum	0.0459	1.4697 (1.3431, 1.6081)

4.3.2 Prevention of tuberculosis

The second example is the meta-analysis of 13 clinical trials to assess the efficacy of Bacillus Calmette-Guérin (BCG) vaccine for the prevention of tuberculosis (TB), which was originally reported by Colditz et al. (1994) and with further details of methodology discussed in Berkey et al. (1995), Sutton et al. (2000), van Houwelingen et al. (2002). We have extracted data on covariates that might explain the heterogeneity among study results from those articles.

The investigators compared two arms. The treatment arm is defined as receiving BCG vaccine, and the control arm as not receiving BCG vaccine. All trials have equivalent surveillance procedures and similar lengths of follow-up among the vaccinated and non vaccinated group. The focus of interest is the occurrence of TB. Latitude is one of several factors which is historically suspected of being associated with the efficacy of BCG vaccine. Latitude represents the variation in rainfall, humidity, environmental mycobacteria that may produce the level of natural immunity against TB, and other factors that may have an influence on the efficacy of BCG vaccine. In the literature, there are a variety of methods of allocation of treatment which could be used as covariates that might explain the heterogeneity of study results. The method of treatment allocation consists of random, alternate, and systematic. However, the 13 reviewed studies have been conducted over a period of more than 60 years, so the year of publication could also be used as one covariate in our analysis.

Therefore, we have applied the modeling of covariate information using the profile likelihood to find out whether distance of each trial from the equator (absolute latitude), direction of latitude from equation, method of treatment allocation, and year of publication are associated with the efficacy of BCG vaccine. The count data and characteristics of the 13 studies on the efficacy of BCG vaccine for the prevention of TB are presented in Table 4.4. It has been determined that *Latitude* is a continuous covariate to describe distance of each trial from the equator; *Direct* is the binary covariate to describe direction of latitude from the equator; North (Direct=0) or South (Direct=1); *Alloc* is the categorical covariate to describe the method of allocation of subjects to BCG vaccine and control groups; random allocation (Alloc=1) or alternate allocation (Alloc=2) or systematic allocation (Alloc=3); and *Year* is continuous covariate to describe the year of publication.

The results of fitting the various profile likelihood models to the meta-analysis of the efficacy of BCG vaccine study are presented in Table 4.5. Also, a forward selection procedure and PLRT have been applied to select the significant covariates. The critical value (5% level) of PLRT with 1 df, which is 3.841, has been used for comparison among possible models. However, the critical value (5% level) of PLRT with 2 df, which is 5.991, has been used for comparison of the model with categorical covariate allocation. The results in Table 4.5

indicate that the only covariate latitude is significantly associated with the efficacy of BCG vaccine for prevention of TB. The estimation of RR and 95% CI are presented in Table 4.6. It is clear that the efficacy of BCG vaccination increases with increasing distance from the equator.

Table 4.4 *Count data and characteristics of 13 studies on the efficacy of BCG vaccine for the prevention of TB*

Trial	x^T	n^T	x^C	n^C	Latitude	Direct	Alloc	Year
1	4	123	11	139	44	0	1	48
2	6	306	29	303	55	0	1	49
3	3	231	11	220	42	0	1	60
4	62	13598	248	12867	52	0	1	77
5	33	5069	47	5808	13	0	2	73
6	180	1541	372	1451	44	0	2	53
7	8	2545	10	629	19	0	1	73
8	505	88391	499	88391	13	0	1	80
9	29	7499	45	7277	27	1	1	68
10	17	1716	65	1665	42	0	3	61
11	186	50634	141	27338	18	0	3	74
12	5	2498	3	2341	33	0	3	69
13	27	16913	29	17854	33	0	3	76

Note. x^T=number of TB cases in the vaccinated group,
n^T=number of persons in the vaccinated group,
x^C=number of TB cases in the unvaccinated group,
n^C=number of persons in the unvaccinated group.

Table 4.5 *Results of fitting various single-covariate models to the multicenter trial of BCG vaccine*

$L^*(\hat{\beta})$	Covariates	$\hat{\beta}_j$	S.E.	P-value	LR-Test
-26636.7200	Intercept	-0.4551	0.0403	0.0000	
-26570.8200^c	Intercept	0.3571	0.0814	0.0000	131.8000^a
	Latitude	-0.0301	0.0027	0.0000	
-26636.7200	Intercept	-0.4547	0.0409	0.0000	0.0000^a
	Direct	-0.0147	0.2416	0.4757	
-26630.0600	Intercept	-0.3530	0.0530	0.0000	13.3200^a
	Alloc-2	-0.3596	0.0995	0.0002	
	Alloc-3	-0.0876	0.1069	0.2064	
-26612.9500	Intercept	-2.3085	0.2787	0.0000	47.5400^a
	Year	0.0260	0.0038	0.0000	
-26568.7200	Intercept	0.3925	0.0857	0.0000	4.2000^b
	Latitude	-0.0331	0.0031	0.0000	
	Alloc-2	0.2112	0.1162	0.0346	
	Alloc-3	-0.0476	0.1105	0.3335	
-26569.7900	Intercept	1.0288	0.4802	0.0161	2.0600^b
	Latitude	-0.0341	0.0039	0.0000	
	Year	-0.0079	0.0056	0.0775	

[a] Comparison of the current model with the model without covariates.

[b] Comparison of the current model with the model with the covariate latitude.

[c] Selected model of the multicenter trial of BCG vaccine.

Table 4.6 *The results of estimation of RR and 95% CI of the meta-analysis of BCG vaccine*

Latitude	S.E.	RR (95% CI)
13	0.0543	0.9659 (0.8684, 1.0743)
18	0.0467	0.8307 (0.7581, 0.9104)
19	0.0455	0.8061 (0.7373, 0.8813)
27	0.0415	0.6334 (0.5839, 0.6871)
33	0.0456	0.5286 (0.4834, 0.5779)
42	0.0599	0.4030 (0.3584, 0.4532)
44	0.0639	0.3794 (0.3348, 0.4300)
52	0.0815	0.2981 (0.2541, 0.3497)
55	0.0886	0.2723 (0.2290, 0.3240)

4.3.3 Ischaemic heart disease

The third example is the multicenter study of 28 trials that study the effect of the average reduction in serum cholesterol on the reduction of the risk of ischaemic heart disease (IHD). These data are taken from Thompson and Sharp (1999). The cholesterol reduction is determined as the reduction in the treated group minus that in the control group, averaged over the follow-up period of the trial. This average extent of cholesterol reduction varied widely across the trials, from 0.3 to 1.5 mmol/l. In these trials, cholesterol was reduced by a variety of interventions. They consist of diets, drugs, and, in one case, surgery. Furthermore, the duration of trials varied widely across the trials, from 0.3 to 12 years. Trial-specific count data and study characteristics of 28 trials are given in Table 4.7. It has been determined that *Chol* is the continuous covariate to describe the reduction in serum cholesterol; *Treat* is the categorical covariate to describe the type of intervention: dietary (Treat=1), drugs (Treat=2), and surgery (Treat=3); and *Time* is the categorical covariate to describe the duration of the trials: less than 2 years (Time=1), between 2.1 and 5 years (Time=2), and between 5.1 and 12 years (Time=3).

The results of fitting the various profile likelihood models to the IHD data are presented in Table 4.8. Also, a forward selection procedure and PLRT have been applied to select the significant covariates. The critical value (5% level) of PLRT with 1 df, which is 3.841, has been used for comparison between the model without covariates and the model with continuous covariate cholesterol. Furthermore, the critical value (5% level) of PLRT with 2 df, which is 5.991, has been used for comparison among the rest of the possible models. The results from Table 4.8 indicate that only cholesterol reduction has a significant effect on the risk of IHD. In addition, the estimate of $\hat{\beta}$ for cholesterol reduction covariate was negative, meaning that the reduction in the risk of IHD actually increases according to the extent of cholesterol reduction. The estimation of RR and 95% CI are presented in Table 4.9.

Table 4.7 *Count data and study characteristics of 28 clinical trials on the serum cholesterol reduction to reduce the risk of IHD*

Trial	x^T	n^T	x^C	n^C	Chol	Treat	Time
1	173	5331	210	5296	0.55	2	3
2	54	244	85	253	0.68	2	2
3	54	350	75	367	0.85	2	2
4	676	2222	936	2789	0.55	2	2
5	42	145	69	284	0.59	2	2
6	73	279	101	276	0.84	2	2
7	157	1906	193	1900	0.65	2	3
8	6	71	11	72	0.85	2	3
9	36	1149	42	1129	0.49	2	2
10	2	88	2	30	0.68	2	1
11	56	2051	84	2030	0.69	2	3
12	1	94	5	94	1.35	2	1
13	131	4541	121	4516	0.70	1	2
14	52	424	65	422	0.87	1	3
15	45	199	52	194	0.95	1	3
16	61	229	81	229	1.13	1	2
17	37	221	24	237	0.31	1	2
18	8	28	11	52	0.61	1	1
19	47	130	50	134	0.57	1	2
20	82	421	125	417	1.43	3	3
21	62	6582	20	1663	1.08	2	1
22	2	94	0	52	1.48	2	1
23	1	23	0	29	0.56	2	1
24	3	60	5	30	1.06	1	2
25	132	1018	144	1015	0.26	1	1
26	35	311	24	317	0.76	2	2
27	3	79	4	78	0.54	2	1
28	7	76	19	79	0.68	2	2

Note. x^T=number of patients with IHD in the treatment group,
n^T=number of patients in the treatment group,
x^C=number of patients with IHD in the control group,
n^C=number of patients in the control group.

Table 4.8 *Results of fitting various models to the multicenter trial of IHD*

$L^*(\hat{\beta})$	Covariates	$\hat{\beta}_j$	S.E.	P-value	LR-Test
-35905.3200	Intercept	-0.1541	0.0299	0.0000	
-35900.6600^c	Intercept	0.0921	0.0861	0.1424	9.3200^a
	Chol	-0.3717	0.1222	0.0012	
-35902.1900	Intercept	-0.0619	0.0613	0.1563	6.2600^a
	Treat-2	-0.1053	0.0707	0.0683	
	Treat-3	-0.3693	0.1548	0.0085	
-35902.4200	Intercept	-0.1152	0.1044	0.1348	5.8000^a
	Time-2	0.0079	0.1109	0.4716	
	Time-3	-0.1515	0.1183	0.1001	
-35899.2600	Intercept	0.1995	0.1241	0.0540	2.8000^b
	Chol	-0.4014	0.1658	0.0077	
	Treat-2	-0.1186	0.0710	0.0474	
	Treat-3	-0.0566	0.2016	0.3894	
-35899.3900	Intercept	0.0290	0.1192	0.4038	2.5400^b
	Chol	-0.3270	0.1327	0.0069	
	Time-2	0.0704	0.1128	0.2663	
	Time-3	-0.0373	0.1259	0.3836	

[a] Comparison of the current model with the model without covariates.

[b] Comparison of the current model with the model with the covariate chol.

[c] Selected model for the multicenter trial of IHD.

Table 4.9 *Results of estimation of RR and 95% CI of the multicenter trials of IHD*

Cholesterol	S.E.	RR (95% CI)
0.26	0.0574	0.9955 (0.8895, 1.1140)
0.31	0.0523	0.9771 (0.8820, 1.0826)
0.49	0.0365	0.9139 (0.8508, 0.9816)
0.54	0.0333	0.8971 (0.8403, 0.9577)
0.55	0.0328	0.8937 (0.8381, 0.9531)
0.56	0.0323	0.8904 (0.8357, 0.9487)
0.57	0.0319	0.8871 (0.8334, 0.9443)
0.59	0.0311	0.8805 (0.8284, 0.9359)
0.61	0.0305	0.8740 (0.8233, 0.9279)
0.65	0.0299	0.8611 (0.8121, 0.9131)
0.68	0.0300	0.8516 (0.8030, 0.9031)
0.69	0.0301	0.8484 (0.7998, 0.8999)
0.70	0.0302	0.8453 (0.7966, 0.8969)
0.76	0.0322	0.8266 (0.7760, 0.8805)
0.84	0.0370	0.8024 (0.7463, 0.8628)
0.85	0.0377	0.7994 (0.7424, 0.8608)
0.87	0.0393	0.7935 (0.7347, 0.8570)
0.95	0.0462	0.7703 (0.7035, 0.8433)
1.06	0.0571	0.7394 (0.6611, 0.8270)
1.08	0.0592	0.7339 (0.6535, 0.8243)
1.13	0.0646	0.7204 (0.6347, 0.8176)
1.35	0.0893	0.6638 (0.5573, 0.7908)
1.43	0.0985	0.6444 (0.5312, 0.7817)
1.48	0.1044	0.6325 (0.5155, 0.7761)

4.4 Summary

To summarize, the modeling of covariate information using the profile likelihood approach becomes attractive in the analysis of meta-analysis of clinical trials as well as in multicenter studies when only covariates on the study level have been considered.

First, the model that has been developed in this chapter, is based upon the Poisson distribution which is appropriate for the structure of binary outcome.

Second, the canonical link has been applied to link the linear predictor to the relative risk which guarantees that the relative risk estimate is positive which is an essential requirement.

Third, the nuisance parameter has been eliminated before dealing with the inference for the parameter of interest, thereby keeping the dimensionality of the approach low. This will lead to more precision in the estimator for the parameter of interest.

Fourth, the software tool that has been developed in this study, is available to compute and deal with this approach.

However, the model that has been developed in this study, does not include the unobserved heterogeneity of the treatment effects in the model. The model for incorporating covariate information with unobserved heterogeneity of the treatment effects is developed in Chapter 6.

CHAPTER 5

Alternative approaches

Besides the profile approach there are other methods to estimate the treatment effect in a MAIPD. A conventional approach is the approximate likelihood (AL) method and another is the multilevel (ML) approach. Both methods are introduced in the next two sections.

5.1 Approximate likelihood model

This approach considers the logarithmic relative risk $\phi_i = \log(\theta_i)$, which can be estimated as

$$\hat{\phi}_i = \log\left(\frac{x_i^T}{n_i^T}\right) - \log\left(\frac{x_i^C}{n_i^C}\right),$$

assuming nonzero events in both arms of trial i. Frequently, it is assumed that - conditional upon study i - the distribution of $\hat{\phi}_i$ might be validly approximated by a normal distribution with unknown mean ϕ_i and known standard deviation σ_i, though it is sufficient to assume that the log-likelihood is well approximated by a quadratic, see van Houwelingen et al. (2002) and Aitkin (1999b).

Therefore, the associated kernel $f(x_i|\theta_j)$ in the mixture model is the normal density, written as

$$f_{AL}(\hat{\phi}_i|\theta_i, \sigma_i^2) = \frac{1}{\sqrt{2\pi}\sigma_i} e^{-\frac{(\hat{\phi}_i - \log(\theta_i))^2}{2\sigma_i^2}}. \tag{5.1}$$

Note that $\hat{\phi}_i$ is the log-relative risk, a scalar, and the individually pooled counts for trial i are not used. In the most general case, the variance is unknown and must be estimated. Under the assumption of Poisson distributed observations, the variance of the log-relative risk in the i−th study can be estimated using the *delta* method (see also Woodward (1999)) as

$$\widehat{var}(\hat{\phi}_i) = \hat{\sigma}_i^2 = \frac{1}{x_i^T} + \frac{1}{x_i^C}, \tag{5.2}$$

so that $\hat{\sigma}_i^2$ is used as a known parameter in (5.1). However, this simple formula has a number of drawbacks. Firstly, (5.2) is not defined if any of the trials has 0 successes. Secondly, for small trials, the normal approximation might not be satisfactory. Thirdly, the variance approximation used in the *delta* method might be too crude to give a good approximation of the true variance of the log-relative risk, especially if the event rate is different from 0.5. Nevertheless, the approach is quite popular, see Whitehead and Whitehead (1991) and Thompson (1993) - likely due to its simplicity and lack of alternatives. To capture heterogeneity, the density (5.1) is used as a kernel in the mixture distribution leading to

$$f(x_i|P) = \sum_{j=1}^{m} f_{AL}(\hat{\phi}_i|\theta_j, \hat{\sigma}_i^2)q_j. \tag{5.3}$$

The AL model was originally named the fixed-effects model by DerSimonian and Laird (1986) and later by Whitehead (2002) for the case of homogeneity to distinguish it from the *random-effects model*. DerSimonian and Laird (1986) suggested to adjust for unobserved heterogeneity. The latter can be viewed as being further modeled with the above discrete mixture following Laird (1978). Here, we prefer to use the term AL model focusing on the fact the normal kernel in the mixture can be viewed as an approximating likelihood. Modeling covariate information for this situation has been discussed in Hedges (1994a), and more generally in Cooper and Hedges (1994), Thompson and Sharp (1999), Brockwell and Gordon (2001), and Böhning (2000) among others. A general introduction is also given by van Houwelingen et al. (2002).

5.2 Multilevel model

The ML model has become very popular in the biometrical literature (see Aitkin (1999a)) and captures the hierarchical structure of the data used in the meta-analysis. The first level (within study) can be modeled by means of the log-linear regression, with

$$\log(p_i^C) = \alpha_i$$
$$\log(p_i^T) = \alpha_i + \beta_i.$$

α_i is in this case the baseline parameter and $\beta_i = \phi_i = \log\left(p_i^T/p_i^C\right)$ is the effect parameter, the log-relative risk.* Under the assumption that the observations are Poisson distributed, the likelihood for trial i is given as

$$f_{ML}(x_i|p_i^C, p_i^T) = e^{-n_i^T p_i^T} \frac{\left(n_i^T p_i^T\right)^{x_i^T}}{x_i^T!} \times e^{-n_i^C p_i^C} \frac{\left(n_i^C p_i^C\right)^{x_i^C}}{x_i^C!}, \tag{5.4}$$

where x_i is the vector $(x_i^T, n_i^T, x_i^C, n_i^C)'$. The rates p_i^C and p_i^T can be replaced by their model associated parameters, namely

$$p_i^C = e^{\alpha_i}$$
$$p_i^T = e^{\alpha_i + \beta_i}.$$

---

* We use for this model the notation β to keep the similarity with the existing literature.

This leads to the following likelihood in trial i

$$f_{ML}(x_i|\alpha_i, \beta_i) = e^{-n_i^T e^{\alpha_i + \beta_i}} \frac{\left(n_i^T e^{\alpha_i + \beta_i}\right)^{x_i^T}}{x_i^T!} \times e^{-n_i^C e^{\alpha_i}} \frac{\left(n_i^C e^{\alpha_i}\right)^{x_i^C}}{x_i^C!}. \qquad (5.5)$$

The second level is modeled by means of a nonparametric mixture distribution - as has been discussed and done previously. The most complex form of heterogeneity is considered, allowing baseline and effect heterogeneity, that is each component in the mixing distribution has its own baseline and effect parameter. The mixture distribution has the form:

$$f(x_i|P) = \sum_{j=1}^{m} f_{ML}(x_i|\alpha_j, \beta_j)q_j, \qquad (5.6)$$

$$\text{with } P = \begin{pmatrix} \alpha_1 & \cdots & \alpha_m \\ \beta_1 & \cdots & \beta_m \\ q_1 & \cdots & q_m \end{pmatrix}.$$

5.3 Comparing profile and approximate likelihood

Here, we elaborate on the similarities and differences between the approximate likelihood developed in Section 5.1 and the profile likelihood of Section 2. Let us consider the *approximate log-likelihood* which is given for the i-th study - up to an additive constant - as

$$AL_i(\phi_i) = -\frac{1}{2}(\phi_i - \hat{\phi}_i)^2/\hat{\sigma}_i^2 \qquad (5.7)$$

where $\hat{\phi}_i = \log\left(\frac{x_i^T}{n_i^T}\right) - \log\left(\frac{x_i^C}{n_i^C}\right)$ is the estimated log-relative risk in the i-th study and $\hat{\sigma}_i^2 = 1/x_i^T + 1/x_i^C$ the associated estimated variance. The *profile log-likelihood* in the i-th study is provided as

$$PL_i(\phi_i) = x_i^T \phi_i - (x_i^T + x_i^C)\log(n_i^T e^{\phi_i} + n_i^C) \qquad (5.8)$$

where again $\phi_i = \log(\theta)$ is the log-relative risk in the i-th study.

5.3.1 The likelihoods for center-specific parameters

Comparing both log-likelihoods for center-specific parameters ϕ_i is in principle identical in doing so for only one center. Therefore, we can drop the index i. We have that

$$PL(\phi) \approx PL(\hat{\phi}) + (\phi - \hat{\phi})PL'(\hat{\phi}) + \frac{1}{2}(\phi - \hat{\phi})^2 PL''(\hat{\phi}) \qquad (5.9)$$

$$= PL(\hat{\phi}) + \frac{1}{2}(\phi - \hat{\phi})^2 PL''(\hat{\phi}) \qquad (5.10)$$

using a second-order Taylor expansion around $\hat{\phi}$ and that $PL'(\hat{\phi}) = 0$. One easily verifies that $PL''(\hat{\phi}) = -\frac{x^T x^T}{x^T + x^C} = -\frac{1}{\hat{\sigma}_i^2}$, showing that

$$PL(\phi) \approx PL(\hat{\phi}) + AL(\phi) \tag{5.11}$$

so that profile and approximate log-likelihood become identical in a neighborhood of the maximum likelihood estimator with both log-likelihoods sharing the same curvature. Of course, both log-likelihoods are maximized by the same estimator.

5.3.2 The likelihoods for restricted parameters

Comparison of log-likelihoods starts to become different when parameters are restricted such as in the situation of the hypothesis of *homogeneity*, for example $\phi_1 = \phi_2... = \phi_k = \phi$. Using independence of the k studies, the *approximate log-likelihood* becomes

$$AL(\phi) = \sum_i AL_i(\phi) = -\frac{1}{2}\sum_i(\phi - \hat{\phi}_i)^2/\hat{\sigma}_i^2, \tag{5.12}$$

and the *profile log-likelihood* takes the form

$$PL(\phi) = \sum_i PL_i(\phi) = \sum_i x_i^T \phi \quad (x_i^T \mid x_i^C)\log(n_i^T e^\phi \mid n_i^C) \tag{5.13}$$

Figure 5.1 shows both log-likelihoods from the example in Table 1.9. Note that $AL(\phi)$ is maximized for $\hat{\phi}_w = \frac{\sum_i \hat{\phi}_i/\hat{\sigma}_i^2}{\sum_i 1/\hat{\sigma}_i^2}$, see van Houwelingen et al. (2002). In addition, the curvature is given as

$$AL''(\phi) = -\sum_i 1/\hat{\sigma}_i^2 = -\sum_i \frac{x_i^T x_i^C}{x_i^T + x_i^C}. \tag{5.14}$$

The profile log-likelihood is maximized for $\hat{\theta} = \exp(\hat{\phi})$ satisfying

$$\hat{\theta} = \frac{\sum_i x_i^T n_i^C/(n_i^T \hat{\theta} + n_i^C)}{\sum_i x_i^C n_i^T/(n_i^T \hat{\theta} + n_i^C)} \tag{5.15}$$

and has curvature

$$PL''(\phi) = \sum_i -(x_i^T + x_i^C)\frac{n_i^T n_i^C e^\phi}{(n_i^T e^\phi + n_i^C)^2}. \tag{5.16}$$

Approximate and profile log-likelihood are not only maximized at different parameter values, the curvature of the profile log-likelihood at the maximum likelihood estimate is different from the curvature of the approximate log-likelihood. To explore this point in more detail, let us assume that the trial is balanced, that is $n_i^T = n_i^C$ for all i. Then, the profile maximum likelihood estimator is available in closed form $\hat{\theta} = \frac{\sum_i x_i^T}{\sum_i x_i^C}$ and the curvature at $\hat{\phi}$ is

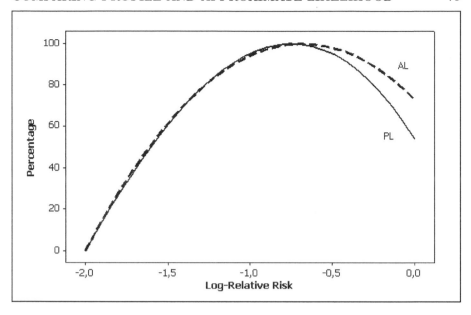

Figure 5.1 *Comparison of the profile and approximate log-likelihood from the meta-analysis Table 5.2*

simply

$$PL''(\hat{\phi}) = \sum_i -(x_i^T + x_i^C) \frac{\frac{\sum_i x_i^T}{\sum_i x_i^C}}{(\frac{\sum_i x_i^T}{\sum_i x_i^C} + 1)^2}$$

$$= -\frac{\sum_i x_i^T \sum_i x_i^C}{\sum_i (x_i^T + x_i^C)}. \qquad (5.17)$$

It is remarkable that a general comparison between these two curvatures is possible, indicating a more precise estimator based upon the profile likelihood.

THEOREM. Let all the centers involved in the MAIPD be balanced, that is $n_i^T = n_i^C$ for all i. Then,

$$PL''(\hat{\phi}) \leq AL''(\phi)$$

and

$$\widehat{var}(\hat{\phi}_{AL}) \leq \widehat{var}(\hat{\phi}_{PL}).$$

Proof. We show

$$\frac{\sum_i x_i^T \sum_i x_i^C}{\sum_i (x_i^T + x_i^C)} \geq \sum_i \frac{x_i^T x_i^C}{x_i^T + x_i^C}$$

Table 5.1 *Results of MAIPD (Table 1.9) from the approximate and profile likelihood model (H is number of studies that belong to the respective component).*

Profile Likelihood Model			Approximate Likelihood Model		
Comp.	1.		Comp.	1.	
θ	.474		θ	.501	
q	1.000		q	1.000	
H	22		H	22	
Log-L.$= -116.606$			Log-L.$= -20.639$		
max $GF = 2.157$			max $GF = 1.389$		
BIC$= -236.304$			BIC$= -44.368$		
Comp.	1.	2.	Comp.	1.	2.
θ	.563	.231	θ	.559	.258
q	.615	.385	q	.678	.322
H	12	10	H	13	9
Log-L.$= -113.360$			Log-L.$= -19.419$		
max $GF = 1.000$			max $GF = 1.000$		
BIC$= -235.992$			BIC$= -48.111$		

or equivalently

$$\frac{\frac{1}{k}\sum_i x_i^T \frac{1}{k}\sum_i x_i^C}{\frac{1}{k}\sum_i(x_i^T + x_i^C)} \geq \frac{1}{k}\sum_i \frac{x_i^T x_i^C}{x_i^T + x_i^C}.$$

This follows from the fact that the function $g(y, z) = \frac{yz}{y+z}$, defined for $y > 0$, $z > 0$ is concave which is proved by showing that the *Hessian* of $g(y, z)$

$$\begin{pmatrix} -2z & y-z \\ z-y & -2y \end{pmatrix} / (y+z)^3$$

is negative definite. This ends the proof.

5.4 Analysis for the MAIPD on selective tract decontamination

In this section our objective is to analyze the relative risk structure of the MAIPD provided in Table (1.9). For this meta-analysis, 22 trials were included to investigate the effect of selective decontamination of the digestive tract on the risk of respiratory tract infection (Selective Decontamination of the Digestive Tract Trialists' Collaborative Group (1993)), described in Section 1.5. The results from the AL and PL models are given in Table (5.1). Both methods give two mixture components as the largest number of components.

The PL classified 12 studies and the AL 13 studies to the first component with an estimated relative risk of 0.56. Consequently, in these studies the risk of respiratory infection is almost halved in comparison to the control group. The second component estimated a relative risk of 0.23 (PL) and 0.25 (AL). Apparently, the estimators from both models are very close together. One difference lies in the BIC. In the fixed-effect model the BIC estimated only one component, whereas in the PL model two components were chosen. In

Table 5.2 *Results of MAIPD (Table 1.9) using the ML model (H is number of studies that belong to the respective component).*

Multilevel Model					
Comp. 1.					
α $\quad$ -1.228					
θ $\quad$ .468					
q $\quad$ 1.000					
H $\quad$ 22					
Log-L.$= -243.586$					
$\max GF = 2.78 \times 10^{12}$					
BIC$= -493.353$					

Comp.	1.	2.			
α	-1.660	$-.660$			
θ	.549	.418			
q	.502	.498			
H	11	11			
Log-L.$= -171.326$					
$\max GF = 6344.888$					
BIC$= -358.106$					

Comp.	1.	2.	3.	4.	
α	-1.577	$-.778$	$-.044$	-1.945	
θ	.635	.379	.508	.230	
q	.274	.414	.091	.222	
H	6	9	2	5	
Log-L.$= -140.834$					
$\max GF = 2.207$					
BIC$= -315.670$					

Comp.	1.	2.	3.	4.	5.	6.
α	-1.593	$-.835$	$-.044$	-1.957	$-.686$	-1.528
θ	.698	.475	.508	.228	.213	.343
q	.224	.202	.091	.212	.213	.058
H	5	3	2	5	6	1
Log-L.$= -138.958$						
$\max GF = 1.000$						
BIC$= -330.463$						

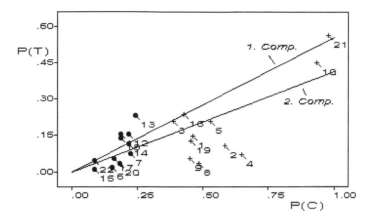

Figure 5.2 *Study allocation to the components for the multilevel model (circle = 1.
comp., cross = 2. comp.).*

the ML many more components were found (Table 5.2). In this approach a
maximum of six components were observed. The fourth and fifth component
estimate nearly the same relative risk, only the baseline is different. With the
BIC, four components were selected as appropriate number of components.
One important difference between the ML model and the PL and AL model is
found in the way the classification of studies into the associated components is
done, see Figure 5.2 and 5.3. The study allocation of the AL model is similar
to the PL model, only the first study is allocated differently. In Figure 5.2, it
can be seen that, for example the third and the 16th study are very close to the
first component line. This means that these studies have the same or similar
relative risk as the first component in the ML model, although the studies
are classified into the second component. The reason for this misallocation
lies in the influence of the baseline heterogeneity on the estimation of effect
heterogeneity. In contrast, Figure 5.3 shows that all studies are allocated on
the basis of the treatment effect in the PL and AL model (which is the major
interest of the practitioner).

5.5 Simulation study

In this section, all three models are compared by means of simulation studies.

5.5.1 Two-component effect and baseline heterogeneity

It is assumed that in the first simulation experiment the population of interest
consists of two clusters. The clusters are represented by the mixing distribu-

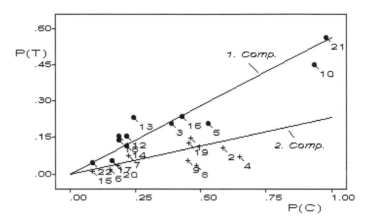

Figure 5.3 *Study allocation to components for the profile likelihood model (circle =
1. comp., cross = 2. comp.).*

tion $P = \begin{pmatrix} 0.5 & 1.5 \\ 0.5 & 0.5 \end{pmatrix}$. Both components receive an identical weight of 0.5.
The first component has a relative risk of 0.5 and the second of 1.5. To mimic
baseline variation, the baseline risks $p_1^C, ...p_k^C$ were generated from a uniform
distribution from 0.1 to 0.66. The parameter p_i^T depends on the component
the i-th study belongs to. If the i-th study belongs to the first component,
then $p_i^T = \theta_1 p_i^C = 0.5 p_i^C$, otherwise $p_i^T = \theta_2 p_i^C = 1.5 p_i^C$. In this case the
weights are equal, so that component membership of each study is generated
by means of a Bernoulli distribution with 0.5 event probability. The sample
size n_i^T and n_i^C were generated from a Poisson distribution with parameter
100. Poisson variates X_i^T with parameters n_i^T and p_i^T and Poisson variates
X_i^C with parameters n_i^C and p_i^C were drawn for each study i, $i = 1, ..., k$. In
this case the number of studies was chosen to be $k = 100$. For reasons of comparability only a two-component mixture was estimated for all three models.
The procedure was replicated 1,000 times. From this replication the mean and
variance of each component were computed. The results of this constellation
are provided in Figure 5.4. The first component of the ML model is considerably overestimated. Note that actually the true relative risk is not captured by
the confidence interval. In contrast the second component is underestimated.
In the other two models the true distribution is discovered.

5.5.2 Two-component effect heterogeneity under baseline homogeneity

The second simulation study shows the extent of influence of the baseline
parameter. All settings have been taken over from the previous simulation
study, besides a $p_i^C = 0.3$ assumption. Now it can be observed in Figure
5.5 that the ML model is much more capable of recovering the true mixture

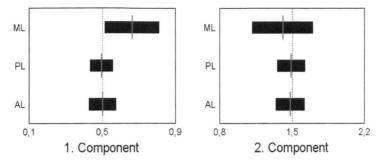

Figure 5.4 *Results of a simulation study of a two-component mixture (with baseline heterogencity) for the three models AL, PL, ML to estimate the predetermined mixing components* {0.5, 1.5} *with weights* {0.5, 0.5}. *The means with 95% CI for each estimated component are displayed.*

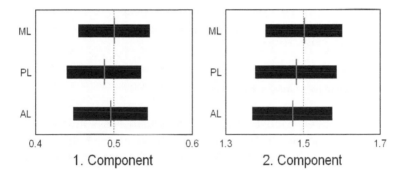

Figure 5.5 *Results of the simulation study of a two-component mixture model (with baseline homogeneity) for the three models AL, PL, ML to estimate the predetermined mixing components* {0.5, 1.5} *with weights* {0.5, 0.5}. *The means with 95% CI of each estimated component are displayed.*

distribution. The confidence intervals from the three models are of the same dimension.

5.5.3 Under effect homogeneity

The next simulation study investigated the situation of effect homogeneity. In this case we used a bootstrap simulation, see Efron (1993). The differences are expected in the sparsity case, where the number of observations and participants are rare. For this we used the sparsity meta-analysis, namely CALGB study adopted from Lipsitz et al. (1998) in Table 1.8. The main settings for the simulation: p_i^C, n_i^T, and n_i^C stem from the sparsity study; θ is predetermined and fixed for all studies $i = 1, ..., k$; X_i^T with parameters $n_i^T \times \theta p_i^C$ and X_i^C with parameters $n_i^C \times p_i^C$ were generated from a Poisson distribution. Note that if any zeros occur in the treatment or control arm, 0.5 is added to each cell. Figure 5.6 shows the bias of 30 values for θ in the interval from 0.1 to 0.99 for all the three models. In this figure the PL-model estimator has the smallest bias. In contrast, the true value is significantly overestimated by the other two models. One reason could be that the estimator of the ML model, here the crude risk ratio estimator

$$\hat{\theta}_{crude} = \frac{\sum_{i=1}^{k} x_i^T \sum_{i=1}^{k} n_i^C}{\sum_{i=1}^{k} n_i^T \sum_{i=1}^{k} x_i^C},$$

does not adjust for a potential center effect. Also, the weighted estimator used in the AL model, where the weight originates from the inverse of the variance of the log-relative risk which might not be appropriate in this situation.

Drawing attention to the variance (Figure 5.7), the AL estimator has a slightly larger value than the ML and PL estimators (which is consistent with the theorem of Section 5.3), whereas the values of the variance of ML and PL models are very close, see Aitkin (1998).

5.6 Discussion of this comparison

One important difference between these three models lies in the way the nuisance or baseline parameter is treated. In the AL model the baseline is integrated into the individual log-relative risk. The specific aspect of the PL approach is that the nuisance parameter is integrated into the likelihood in such a way that the occurring likelihood, the profile likelihood, depends only on the parameter of interest. In contrast, the ML method does not eliminate the nuisance parameter, but estimates it as a separate parameter. The results in Table 5.2 and the simulation study in Figure 5.4 show that this model loses power when estimating baseline heterogeneity. Furthermore, the allocation of studies or centers to the mixed components is also dependent on the baseline parameter. In other words, the baseline parameter has a very strong influence on estimating the treatment effect. In the situation of an increased baseline heterogeneity it can happen, like in the simulation study in Figure 5.4, that the result of estimating the treatment effect heterogeneity is confounded by the existing strong baseline heterogeneity. Consequently, a substantial disadvantage of the ML model can be seen in the handling of the baseline parameter.

It should be mentioned that the profile method is a conventional way to deal with nuisance parameters, but by no means the only way. In the ideal case, see Pawitan (2001), parameter of interest and nuisance parameter are orthogonal, that is, the joint likelihood $\pounds(\theta, p^C) = \pounds_1(\theta)\pounds_2(p^C)$ factors into likelihood depending only on θ and p^C, respectively. For the ease of discussion

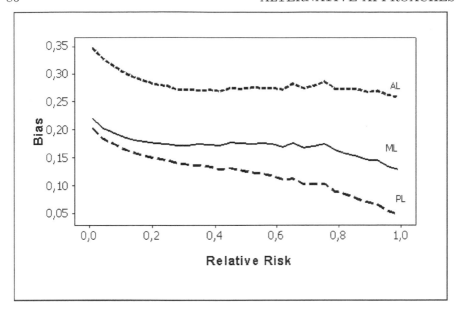

Figure 5.6 *The bias of ML, PL and AL models in a simulation study under effect homogeneity*

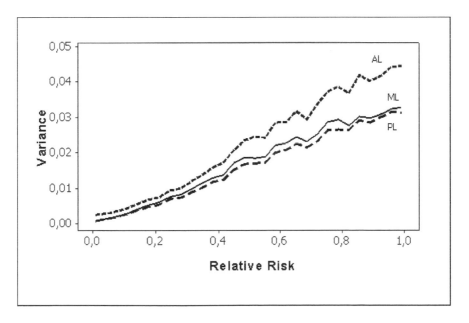

Figure 5.7 *The variance of ML, PL and AL model estimators in a simulation study under effect homogeneity*

only one trial is considered, though generalizations are straightforward. Write the joint likelihood $\exp(-p^T n^T)(p^T n^T)^{x^T} \times \exp(-p^C n^C)(p^C n^C)^{x^C}$ as product of $\mathcal{L}_1(\theta) = \left(\frac{n^T \theta}{n^C + n^T \theta}\right)^{x^T} \left(\frac{n^C}{n^C + n^T \theta}\right)^{x^C}$ and $\mathcal{L}_2(\eta_n) = \exp(-\eta_n)\eta_n^{x^T + x^C}$, where θ is the risk ratio and $\eta_n = n^T p^T + n^C p^C$. In case that the trial is balanced $\eta_n = n^T p^T + n^C p^C = \eta(n^T + n^C)$, and θ and $\eta = p^T + p^C$ are *orthogonal*. In the case of orthogonality, one can solely base inference on $\mathcal{L}_1(\theta)$, and the profile likelihood is identical to $\mathcal{L}_1(\theta)$ which is also a true likelihood. If the trial is unbalanced, the transformation $\eta_n = n^T p^T + n^C p^C$ necessarily incorporates the known, trial-specific sample size parameters, but $\mathcal{L}_1(\theta)$ will remain identical. Alternatively, one may base inference on the likelihood conditional on the sufficient statistic $x = x^T + x^C$ for the nuisance parameter, and, although this is by no means in generality the case, it does turn out again to be $\mathcal{L}_1(\eta)$, see for a more general discussion Pawitan (2001) or McCullagh and Nelder (1989). Yet, another way in dealing with the nuisance parameter is suggested in van Houwelingen et al. (1993). It is suggested to use as effect measure the odds ratio and base inference on the distribution of X^C - conditional upon the margins being fixed. The occurring noncentral hypergeometric distribution is a function of the odds ratio only, so that the associated likelihood is free of the nuisance parameter. This appears to be an attractive approach and - despite the complex character of the noncentral hypergeometric likelihood - should be analyzed in further depth and compared with the profile approach in future work.

5.7 Binomial profile likelihood

One clear alternative to the Poisson profile likelihood is the binomial profile likelihood. In some MAIPD it is realistic to assume a binomial distribution of the observations, especially when n^T and n^C are numbers of patients. The profile likelihood framework is analog applicable for the binomial likelihood function like the Poisson likelihood. The binomial product likelihood over all trials becomes

$$\prod_{i=1}^{k} \binom{n_i^T}{x_i^T} \left(p_i^T\right)^{x_i^T} \left(1 - p_i^T\right)^{n_i^T - x_i^T} \times \binom{n_i^C}{x_i^C} \left(p_i^C\right)^{x_i^C} \left(1 - p_i^C\right)^{n_i^C - x_i^C}$$

and taking the logarithm the *log-likelihood* (ignoring the only data-dependent term) takes the form

$$\sum_{i=1}^{k} x_i^T \log\left(p_i^T\right) + \left(n_i^T - x_i^T\right) \log(1 - p_i^T)$$
$$+ x_i^C \log\left(p_i^C\right) + \left(n_i^C - x_i^C\right) \log(1 - p_i^C). \tag{5.18}$$

5.7.1 Estimation of relative risk in meta-analytic studies using the binomial profile likelihood

In the same way as in Section 2.2 p_i^T is rewritten as $\theta_i p_i^C$. This leads to

$$\sum_{i=1}^{k} x_i^T \log(\theta_i) + (x_i^T + x_i^C) \log(p_i^C)$$
$$+ (n_i^T - x_i^T) \log(1 - \theta_i p_i^C) + (n_i^C - x_i^C) \log(1 - p_i^C) \qquad (5.19)$$

and the binomial profile likelihood as function of $\mathbf{p}^C$ for arbitrary, but fixed $\theta = (\theta_1, ..., \theta_k)'$ is

$$L_{BIN}(\mathbf{p}^C | \theta) = \sum_{i=1}^{k} x_i^T \log(\theta_i) + (x_i^T + x_i^C) \log(p_i^C)$$
$$+ (n_i^T - x_i^T) \log(1 - \theta_i p_i^C) + (n_i^C - x_i^C) \log(1 - p_i^C). \qquad (5.20)$$

The $\mathbf{p}^C$ that maximizes (5.20) can be found by solving the partial derivatives, equated to zero, as follows:

$$\frac{\partial}{\partial \mathbf{p}^C} L_{BIN}(\mathbf{p}^C | \theta) = \sum_{i=1}^{k} \frac{x_i^T + x_i^C}{p_i^C} + \frac{n_i^C - x_i^C}{1 - p_i^C} + \frac{(n_i^T - x_i^T) \theta_i}{1 - \theta_i p_i^C} \qquad (5.21)$$

The solution $\mathbf{p}^C$ - found by setting (5.21) to zero - is by far not as easy as in the case of the Poisson profile likelihood. Here, two solutions are possible:

$$(\hat{p}_i^C)_1 = \frac{n_i^C + n_i^T \theta + \theta x_i^C + x_i^T - \sqrt{(n_i^C + \theta(n_i^T + x_i^C) + x_i^T)^2 - 4\theta x_i n_i}}{2\theta n_i}$$
$$:= p_i^C(\theta)_1 \qquad (5.22)$$

$$(\hat{p}_i^C)_2 = \frac{n_i^C + n_i^T \theta + \theta x_i^C + x_i^T + \sqrt{(n_i^C + \theta(n_i^T + x_i^C) + x_i^T)^2 - 4\theta x_i n_i}}{2\theta n_i}$$
$$:= p_i^C(\theta)_2 \qquad (5.23)$$

with $n_i = n_i^C + n_i^T$ and $x_i = x_i^C + x_i^T$

Now, these two solutions may be understood as a function of θ. If $\theta = 1$, then (5.22) is equal $\frac{x_i}{n_i}$. In contrast (5.23) is equal 1, a nonfeasible solution in this situation, because the term $\log(1 - \theta_i p_i^C)$ and $\log(1 - p_i^C)$ in the likelihood function (5.20) attain $-\infty$. Consequently, (5.23) can be excluded from this analysis. In the following only (5.22), denoted simply as $p_i^C(\theta)$, is considered. To proceed further, the profile likelihood procedure replaces $\mathbf{p}^C$ in (5.19) by $\mathbf{p}^C(\theta)$. This leads to

$$L_{BIN}(\theta) = \sum_{i=1}^{k} x_i^T \log(\theta_i) + (n_i^T - x_i^T) \log(1 - \theta_i p_i^C(\theta_i))$$
$$+ (x_i^T + x_i^C) \log(p_i^C(\theta_i)) + (n_i^C - x_i^C) \log(1 - p_i^C(\theta_i)). \qquad (5.24)$$

In comparison to the Poisson profile log-likelihood (5.24) is more complicated.

To avoid the division by zero we have to exclude three scenarios. These are $p_i^C(\theta) = 0$, $1 - p_i^C(\theta) = 0$, and $1 - \theta p_i^C(\theta) = 0$. The first case occurs if $x^T = 0$ and $x^C = 0$, then we have

$$p_i^C(\theta) = \frac{n_i^C + n_i^T\theta - \sqrt{\left(n_i^C + \theta n_i^T\right)^2}}{2\theta n_i}$$

$$= 0.$$

The second case occurs if $x^C = n^C$. In this case, it is

$$p_i^C(\theta) = \frac{n_i^C + n_i^T\theta + \theta n_i^C + x_i^T - \sqrt{\left(n_i^C(-1+\theta) + \theta n_i^T - x_i^T\right)^2}}{2\theta n_i}.$$

Here we have to distinguish between $n_i^C(-1+\theta) + \theta n_i^T - x_i^T \geq 0$ and $n_i^C(-1+\theta) + \theta n_i^T - x_i^T < 0$. If $n_i^C(-1+\theta) + \theta n_i^T - x_i^T \geq 0$, then

$$p_i^C(\theta) = \frac{n_i^C + n_i^T\theta + \theta n_i^C + x_i^T - n_i^C(-1+\theta) - \theta n_i^T + x_i^T}{2\theta n_i}$$

$$= \frac{n_i^C + x_i^T}{\theta(n_i^T + n_i^C)}$$

and if $n_i^C(-1+\theta) + \theta n_i^T - x_i^T < 0$, then

$$p_i^C(\theta) = \frac{\theta n_i^T + \theta n_i^C + \theta n_i^C + \theta n_i^T}{2\theta n_i} = 1.$$

The last case occurs if $x^T = n^T$. In this case, we have

$$\theta p_i^C(\theta) = \theta\left(\frac{n_i^C + n_i^T\theta + \theta x_i^C + n_i^T - \sqrt{\left(n_i^C - \theta\left(n_i^T + x_i^C\right) + n_i^T\right)^2}}{2\theta n_i}\right)$$

Again we have to distinguish between $n_i^C - \theta\left(n_i^T + x_i^C\right) + n_i^T \geq 0$ and $n_i^C - \theta\left(n_i^T + x_i^C\right) + n_i^T < 0$. If $n_i^C - \theta\left(n_i^T + x_i^C\right) + n_i^T \geq 0$, then

$$\theta p_i^C(\theta) = \theta\left(\frac{n_i^T + x_i^C}{n_i^T + n_i^C}\right)$$

and if $n_i^C - \theta\left(n_i^T + x_i^C\right) + n_i^T < 0$, then

$$\theta p_i^C(\theta) = \theta\left(\frac{n_i^C + n_i^T + n_i^C + n_i^T}{2\theta n_i}\right) = 1.$$

Consequently, in the case of x^T and x^C being both equal to zero, or if $x^T = n^T$

or $x^C = n^C$ we add 0.5 to x^T or x^C and add 1 to n^T or n^C, respectively. In comparison to the Poisson profile likelihood it was not necessary here to consider (and exclude) the possibility of dividing by zero.

5.7.2 The binomial profile likelihood under effect homogeneity

In this section we assume homogeneity in the effect parameter, with $\theta_1 = \theta_2 = \ldots = \theta_k$. To accomplish maximum likelihood estimation we consider the derivative of (5.24). In this case the derivative is more complex, but it can be written as

$$
\frac{\partial}{\partial \theta} L_{BIN}(\theta) = \sum_{i=1}^{k} \frac{x_i^T}{\theta} + \frac{\left(x_i^T + x_i^C\right)\left(p_i^C\right)'(\theta)}{p_i^C(\theta)} - \frac{\left(n_i^C - x_i^C\right)\left(p_i^C\right)'(\theta)}{1 - p_i^C(\theta)}
$$
$$
- \frac{(n_i^T - x_i^T)(p_i^C(\theta) + \theta\left(p_i^C\right)'(\theta))}{1 - \theta p_i^C(\theta)}
\tag{5.25}
$$

In (5.25) we write only $\left(p_i^C\right)'(\theta)$ as a function of θ. The interested reader may find full details on derivative calculations in the appendix A.1. No closed form solution for $\hat\theta$ is available. Possible is an iterative construction such as

$$
\theta_{PMLE} = \frac{\sum_{i=1}^{k} x_i^T}{t_i(\theta_{PMLE})}
\tag{5.26}
$$

$$
t_i(\theta) = \sum_{i=1}^{k} \frac{\left(n_i^C - x_i^C\right)\left(p_i^C\right)'(\theta)}{1 - p_i^C(\theta)}
$$
$$
+ \frac{(n_i^T - x_i^T)(p_i^C(\theta) + \theta\left(p_i^C\right)'(\theta))}{1 - \theta p_i^C(\theta)}
$$
$$
- \frac{\left(x_i^T + x_i^C\right)\left(p_i^C\right)'(\theta)}{p_i^C(\theta)}
$$

The first disadvantage is that the same proof of existence, uniqueness of the solution is *not* available as it was achieved for the iteration (2.22) in Section 2.4. The second disadvantage is the increased complexity of this approach.

5.7.3 Comparison of the Poisson and binomial likelihood in an example

For this comparison we take the beta-blocker (Table 1.5) study as an example. The point estimate of the relative risk is 0.791191 for the Poisson and 0.790892 for binomial profile likelihood. Both estimates are very close together as the log-likelihood function of binomial and Poisson profile likelihood in Figure 5.8 indicate.

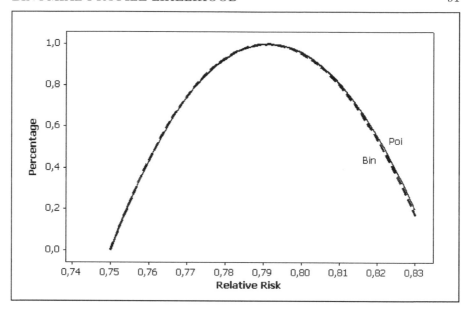

Figure 5.8 *Comparison of the binomial and Poisson profile likelihood from the beta-blocker study (Table 1.5)*

In the sparsity case it is frequently necessary to add 0.5 when using the binomial profile likelihood. If we compare then both likelihoods, the differences are becoming more substantial than before (see Figure 5.9 for illustration). Note that we don't need to add 0.5 to the data when using the Poisson profile likelihood, so that the differences notable in Figure 5.9 are mostly due to this addition of 0.5. Again, both likelihoods become more similar when the same constant 0.5 is also added to the data when constructing the Poisson likelihood.

It is also possible for the binomial profile likelihood to include unobserved heterogeneity and observed heterogeneity in the form of covariate information into the modeling. However, this seems not necessary at this stage, because the Poisson profile likelihood can be considered as an appropriated approximation of the binomial profile likelihood. For the sake of completeness we provide an estimation of the variance in the next section.

5.7.4 Variance estimate of the PMLE

Similarly to Section 2.6.1 the variance of the maximum likelihood estimator can be approximated by the negative inverse of the second derivative of the profile log-likelihood function. In this case the second derivative is rather

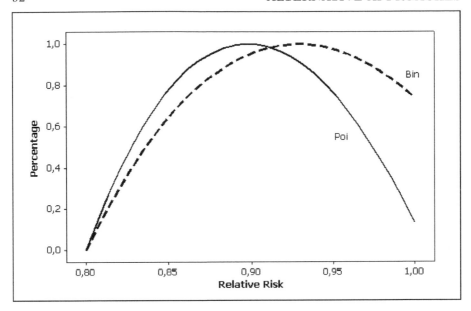

Figure 5.9 *Comparison of the binomial and Poisson profile likelihood from the cancer and leukemia group data (1.8)*

complex and only for completeness given here as

$$
\frac{\partial^2}{\partial \theta^2} L_{BIN}(\theta) = \sum_{i=1}^{k} \frac{x_i^T}{\theta^2} + (x_i^T + x_i^C) \left(-\frac{\left(p_i^C\right)'(\theta)^2}{p_i^C(\theta)^2} + \frac{\left(p_i^C\right)''(\theta)}{p_i^C(\theta)} \right)
$$
$$
- (n_i^C - x_i^C) \left(\frac{\left(p_i^C\right)'(\theta)^2}{\left(1 - p_i^C(\theta)\right)^2} + \frac{\left(p_i^C\right)''(\theta)}{1 - p_i^C(\theta)} \right)
$$
$$
- (n_i^T - x_i^T) \left(\frac{\left(-p_i^C(\theta) - \theta \left(p_i^C\right)'(\theta)\right)^2}{\left(1 - \theta p_i^C(\theta)\right)^2} - \frac{-2 \left(p_i^C\right)'(\theta) - \theta \left(p_i^C\right)''(\theta)}{1 - \theta p_i^C(\theta)} \right).
$$
$$(5.27)$$

Details on the derivatives of $\left(p_i^C\right)''(\theta)$ are placed in the appendix A.1.

CHAPTER 6

Incorporating covariate information and unobserved heterogeneity

6.1 The model for observed and unobserved covariates

This chapter brings both models of unobserved heterogeneity (see Chapter 3) and observed heterogeneity in the form of covariate information (see Chapter 4) together. Let us start with the common nonparametric mixture distribution given as

$$f_i(Q) = \sum_{j=1}^{m} f_i(\theta_j) q_j \tag{6.1}$$

where $f_i(\theta_j)$ and Q are as defined in formula (3.8) in Chapter 3. On the other hand, we have modeled available covariate information in Chapter 4 by means of a log-linear model

$$\log \theta = \eta = \beta_0 + \beta_1 z_1 + ... + \beta_p z_p,$$

where $z_1, \cdots, z_p$ are covariates expressing information on the study level such as the date of study, the treatment modification, or the location of study. These form via $\eta = \beta_0 + \beta_1 z_1 + ... + \beta_p z_p$ the *linear predictor*. It is clear from (6.1) that in the case of m subpopulations with subpopulation-specific relative risks θ_j, we will have m equations linking the mean structure to the linear predictor:

$$\log \theta_j = \eta_j = \beta_{0j} + \beta_{1j} z_1 + ... + \beta_{pj} z_p. \tag{6.2}$$

Note that there are now m coefficient vectors $(\beta_{0j}, \cdots, \beta_{pj})$ which create some complexities in making choices for them.

To illustrate, let us look at the simple example of one covariate z and $m = 2$. The most general form is

$$\log \theta_1 = \eta_1 = \beta_{01} + \beta_{11} z \tag{6.3}$$

$$\log \theta_2 = \eta_2 = \beta_{02} + \beta_{12} z \tag{6.4}$$

where there are no restrictions on the parameter space and two straight lines are fitted to the two component populations. This is illustrated in Figure 6.1. Alternatively, one might consider models with restrictions on the parameters.

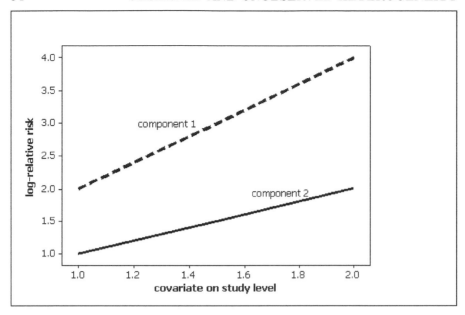

Figure 6.1 *Illustration of a mixture of two log-linear regression models with free intercept and slope*

A model that would be quite meaningful since it has only one common parameter for the covariate effect is

$$\log \theta_1 = \eta_1 = \beta_{01} + \beta_1 z \tag{6.5}$$
$$\log \theta_2 = \eta_2 = \beta_{02} + \beta_1 z \tag{6.6}$$

An illustration is provided in Figure 6.2.

Let us consider the more general situation firstly. For study i we have that

$$\eta_{ij} = \beta_{0j} + \beta_{1j} z_{i1} + \dots + \beta_{pj} z_{ip}.$$

Since θ_j is linked via $\log(\theta_j) = \eta_j = \beta_{0j} + \beta_{1j} z_1 + \dots + \beta_{pj} z_p$, the discrete mixing distribution Q is now giving weights q_j to coefficient vectors $\beta_j = (\beta_{0j}, \beta_{1j}, \cdots, \beta_{pj})^T$. This leads to the discrete probability distribution

$$Q(B) = \begin{pmatrix} \beta_1 & \dots & \beta_m \\ q_1 & \dots & q_m \end{pmatrix} \tag{6.7}$$

$$\text{and } B = \begin{pmatrix} \beta_{01} & \dots & \beta_{p1} \\ \dots & \dots & \dots \\ \beta_{0m} & \dots & \beta_{pm} \end{pmatrix} \tag{6.8}$$

Note that we allow here different $\beta_1, \dots, \beta_p$ for each mixture component $j = 1, \dots, m$.

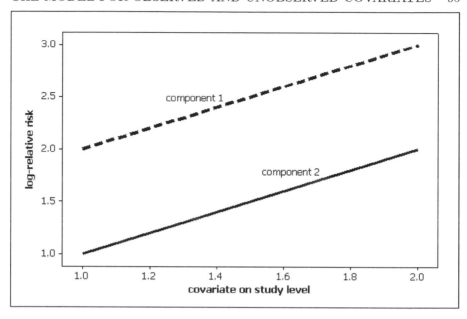

Figure 6.2 *Illustration of a mixture of two log-linear regression models with free intercept and common slope*

The likelihood function over all centers is given as

$$L(Q(B)) = \prod_{i=1}^{k} \sum_{j=1}^{m} f_i(\exp(\eta_{ij}))q_j. \tag{6.9}$$

6.1.1 Finding the maximum likelihood estimator

One way to find the maximum likelihood estimator is accomplished by using the EM algorithm with the Newton-Raphson algorithm nested in the *M-step* of the EM algorithm. To be more specific, let us start with the complete data likelihood as

$$L(Q(B)) = \prod_{i=1}^{k} \prod_{j=1}^{m} (f_i(\exp(\eta_{ij}))q_j)^{y_{ij}}, \tag{6.10}$$

where $y_{ij} = 1$, if center i belongs to subpopulation j, and 0 otherwise. To proceed as in Chapter 3 we replace the unobserved y_{ij} by their expected values as the E-step

$$e_{ij} = \frac{f_i(\exp(\eta_{ij}))q_j}{\sum_{j=1}^{m} f_i(\exp(\eta_{ij}))q_j}. \tag{6.11}$$

This leads to the expected complete data likelihood

$$L(Q(B)) = \prod_{i=1}^{k} \prod_{j=1}^{m} (f_i(\exp(\eta_{ij}))q_j)^{e_{ij}}. \tag{6.12}$$

Maximizing with respect to q_j is easily done to provide

$$\hat{q}_j = \frac{1}{k} \sum_{i=1}^{k} e_{ij}.$$

More challenging is the maximization with respect to B. The log-likelihood function is

$$L(Q(B)) = \sum_{i=1}^{k} \sum_{j=1}^{m} e_{ij} \log f_i(\exp(\eta_{ij})) + e_{ij} \log(q_j)$$

$$= \sum_{i=1}^{k} \sum_{j=1}^{m} e_{ij} \left(x^T \eta_{ij} - (x_i^C + x_i^T) \log(n_i^C + \exp(\eta_{ij})n_i^T) \right) + e_{ij} \log(q_j)$$

$$= \sum_{j=1}^{m} \sum_{i=1}^{k} e_{ij} \left(x^T \eta_{ij} - (x_i^C + x_i^T) \log(n_i^C + \exp(\eta_{ij})n_i^T) \right) + e_{ij} \log(q_j). \tag{6.13}$$

Maximization can be completed by noting that β_j occurs only in $L(\beta_j) = \sum_{i=1}^{k} e_{ij} \left(x^T \eta_{ij} - (x_i^C + x_i^T) \log(n_i^C + \exp(\eta_{ij})n_i^T) \right)$, so that $L(Q(B))$ is maximized by doing m maximizations of $L(\beta_j)$ for $j = 1, ..., m$.

The gradient $\nabla L(\beta_j)$ for the j-th component has elements

$$\frac{\partial}{\partial \beta_{lj}} L(\beta_j) = \sum_{i=1}^{k} e_{ij} \left(x^T z_{il} - (x_i^C + x_i^T)n_i^T \frac{\exp(\eta_{ij})}{n_i^C + \exp(\eta_{ij})n_i^T} z_{il} \right),$$

with the gradient being

$$\nabla L(\beta_j) = \left(\frac{\partial}{\partial \beta_{1j}} L(Q(B)), ..., \frac{\partial}{\partial \beta_{pj}} L(Q(B)) \right)'. \tag{6.14}$$

For each component the *Hesse* matrix has the form

$$\nabla^2 L(\beta_j) = \left(\frac{\partial^2}{\partial \beta_{lj} \partial \beta_{hj}} L(\beta_j) \right) = -\sum_{i=1}^{k} e_{ij} \left(\frac{(x_i^C + x_i^T)n_i^T \exp(\eta_{ij})}{\left(n_i^C + \exp(\eta_{ij})n_i^T \right)^2} z_{il} z_{ih} \right).$$

In matrix notation

$$\nabla^2 L(\beta_j) = -Z'W^*(\beta_j)Z \tag{6.15}$$

where Z is the design matrix containing the information on the p covariates and $W^*(\beta_j)$ is a diagonal matrix defined as

$$W^*(\beta_j) = \begin{pmatrix} w_1(\beta_j) & 0 & \ldots & 0 & 0 \\ 0 & w_2(\beta_j) & \ldots & 0 & 0 \\ \ldots & \ldots & \ldots & \ldots & \ldots \\ 0 & 0 & \ldots & w_{p-1}(\beta_j & 0 \\ 0 & 0 & \ldots & 0 & w_p(\beta_j) \end{pmatrix}$$

with

$$w_i(\beta_j) = e_{ij} \frac{(x_i^C + x_i^T) n_i^T \exp(\eta_{ij})}{\left(n_i^C + \exp(\eta_{ij}) n_i^T\right)^2}.$$

The Newton-Raphson procedure is applicable for each component, as

$$\beta_j^{(n+1)} = \beta_j^{(n)} - \nabla^2 L(\beta_j^{(n)})^{-1} \nabla L(\beta_j^{(n)}). \tag{6.16}$$

This completes the M-step in the EM algorithm.

Additionally, we can use e_{ij} for classification of the centers using the MAP rule (see also Section 3.5).

6.1.2 Improving computational efficiency and reliable convergence

The convergence of the sequence (6.16) is not guaranteed. We would like to point out an important possible simple modification called the *lower bound* method which could improve the algorithmic situation considerably. The lower bound method has at least two important properties:

- reliable convergence toward the maximum likelihood estimate

- savings in terms of computational effort.

Since in this case the weights, which occur in the second derivative matrix, have the property that

$$w_i(\beta_j) = e_{ij} \frac{(x_i^C + x_i^T) n_i^T \exp(\eta_{ij})}{\left(n_i^C + \exp(\eta_{ij}) n_i^T\right)^2} \leq e_{ij}(x_i^T + x_i^C)\frac{1}{4}$$

for all values of η_{ij} (see also the appendix for a proof of this inequality), it is possible to replace the Newton-Raphson step by the lower bound procedure (Böhning and Lindsay (1988), Lange (2004)). The key idea here is to use a global matrix bound for the Hesse matrix if it is available. Suppose that $\nabla^2 L(\beta_j) \geq \mathbf{B}_j$ for all β_j, where $\mathbf{B}_j$ is a negative definite matrix,* then it can be verified that the iteration

$$\beta_j^{(n+1)} = \beta_j^{(n)} + \mathbf{B}_j^{-1} \nabla L(\beta_j^{(n)})$$

* Here $\mathbf{A} \geq \mathbf{B}$ means that $\mathbf{A} - \mathbf{B}$ is nonnegative definite matrix.

is gradually decreasing in the sense that $L(\beta_j^{(n+1)}) \geq L(\beta_j^{(n)})$ where $L(\beta_j)$ is the expected, complete log-likelihood.

$$\beta^{(n+1)} = \beta^{(n)} + \mathbf{B}_j^{-1} \nabla L(\beta^{(n)}) \qquad (6.17)$$

Note that in the equation (6.17) $\mathbf{B}_j$ can be written as $Z'\Lambda_j Z$, where Λ_j is a diagonal matrix with $\Lambda_{j_{ii}} = e_{ij}(x_i^T + x_i^C)\frac{1}{4}$, independent of η. $\mathbf{B}_j$ represents a global upper bound for $-\nabla^2 L(\beta_j)$, e.g. $-\nabla^2 L(\beta_j) \leq \mathbf{B}_j$ for all β_j, where "$\leq$" denotes the matrix ordering. Also it is guaranteed to convergence to the maximum. The equation (6.17) has the advantage to need the global bound matrix to be inverted only once (for details see Böhning and Lindsay (1988), Böhning (1992)). Suppose that n_0 is the average number of iterations required in the Newton-Raphson iteration, the lower bound procedure will save about $n_0 - 1$ matrix inversions.

6.1.3 Finding the standard errors of estimates

In this situation it is a larger challenge to estimate the variance of the maximum likelihood estimators of β_j although it is possible. In fact, it can be seen as one of the benefits of the profile likelihood method to work with a reduced number of parameters, so that the dimensionality of the problem remains low. We note in passing that correct approximations of the standard errors are not obtained as a byproduct of the estimation process as in Section 4.2.3, because the expected, complete likelihood function (6.12) usually has a different curvature in comparison to the observed log-likelihood and is therefore not the appropriate basis for developing estimates of the standard errors.

We will briefly describe two general ways for constructing an approximation of the (expected) information matrix $i(\psi)$ with elements $-E\left(\frac{\partial^2}{\partial \psi_{j1} \partial \psi_{j2}} LL(\psi)\right)$ where ψ is now a vector of size $m \times p + m - 1$ having firstly $m \times p$ elements through the vector of regression coefficients

$$\underline{\beta} = (\beta_1, ..., \beta_m)' \text{ with } \beta_j = (\beta_{0j}, ..., \beta_{pj})'$$

and finally $m - 1$ free parameters through the weights $q_1, ..., q_{m-1}$. If $p = 3$ and $m = 4$ there are 12 β-parameters and 3 weight parameters involved and the information matrix would have a dimension of 15.

- McLachlan and Peel (2000) give an interesting approximation of the information matrix using the gradient only. Consider

$$LL(\psi) = \sum_{i=1}^{k} \log \sum_{j=1}^{m} f_i(\exp(\eta_{ij})) q_j$$

which we write as

$$LL(\psi) = \sum_{i=1}^{k} \log L_i(\psi).$$

Now,

$$\nabla \log L(\psi) \nabla \log L(\psi)' = \sum_{i=1}^{k} \nabla \log L_i(\psi) \nabla \log L_i(\psi)'$$

provides an approximation of $i(\psi)$ since

$$E\left(\sum_{i=1}^{k} \nabla \log L_i(\psi) \nabla \log L_i(\psi)'\right) = i(\psi).$$

A proof of this result is given in McLachlan and Peel (2000). Note that $\sum_{i=1}^{k} \nabla \log L_i(\psi) \nabla \log L_i(\psi)'$ is nonnegative definite, since

$$\mathbf{h}'\left(\sum_{i=1}^{k} \nabla \log L_i(\psi) \nabla \log L_i(\psi)'\right)\mathbf{h} = \sum_{i=1}^{k} (\nabla \log L_i(\psi)'\mathbf{h})^2 \geq 0$$

for any $\mathbf{h} \neq 0$.

- A second approach would more directly consider the *observed* information matrix, namely the *Hesse* matrix of $LL(\psi)$

$$I(\psi) = -\left(\frac{\partial^2}{\partial \psi_{j1} \partial \psi_{j2}} LL(\psi)\right).$$

Let us now consider the observed mixture log-likelihood

$$LL(\psi) = \sum_{i=1}^{k} \log \sum_{j=1}^{m} f_i(\exp(\eta_{ij})) q_j, \tag{6.18}$$

where we take only the logarithm of (6.9). The first partial derivative with respect to $\beta_{j,d1}$ is given as

$$\frac{\partial}{\partial \beta_{j,d1}} \log L_i(\psi) = \frac{\frac{\partial}{\partial \beta_{j,d1}} f_i(\exp(\eta_{ij})) q_j}{\sum_{j^*=1}^{m} f_i(\exp(\eta_{ij^*})) q_{j^*}}, \tag{6.19}$$

where

$$f_i(\exp(\eta_{ij})) = \frac{\exp(\eta_{ij})^{x_i^T}}{\left(n_i^C + n_i^T \exp(\eta_{ij})\right)^{x_i^T + x_i^C}} q_j$$

and

$$\frac{\partial}{\partial \beta_{j,d1}} f_i(\exp(\eta_{ij})) q_j = \frac{z_{id1} \exp(\eta_{ij})^{x_i^T} (-\exp(\eta_{ij}) n_i^T x_i^C + n_i^C x_i^T)}{\left(n_i^C + n_i^T \exp(\eta_{ij})\right)^{x_i^T + x_i^C + 1}} q_j.$$

For example, in establishing the partial derivative with respect to the weight q_j we have to keep in mind that $\sum_{j=1}^{m} q_j = 1$, so that $q_m = 1 - \sum_{j=1}^{m-1} q_j$. Hence, the log-likelihood (6.18) can be equivalently written as

$$\sum_{i=1}^{k} \log \left(\sum_{j=1}^{m-1} \{f_i(\exp(\eta_{ij})) - f_i(\exp(\eta_{im}))\} q_j\right) = \sum_{i=1}^{k} \log L_i(\psi),$$

Table 6.1 *Results of the analysis of the effect of nicotine replacement therapy (NRT) on quitting smoking without consideration of covariate information (H is the number of trials that belong to the respective component).*

Comp.	1.	2.	3.
θ_j	1.760	1.301	5.316
95% CI	(1.446 - 2.142)	(1.032 - 1.640)	(1.154 - 24.490)
q_j	0.6418	0.3342	0.0240
H	52	7	0

Log-L.$= -17,216.9$
BIC $= -34,454.1$
$\max GF = 1.000000$

so that the partial derivative with respect to q_j is simply

$$\frac{\partial}{\partial q_j} \log L_i(\psi) = \frac{f_i(\exp(\eta_{ij})) - f_i(\exp(\eta_{im}))}{\sum_{j^*=1}^{m} f_i(\exp(\eta_{ij^*}))q_{j^*}}. \tag{6.20}$$

The diagonal elements of the inverse of $\sum_{i=1}^{k} \nabla \log L_i(\psi) \nabla \log L_i(\psi)'$ provide estimates for the variances of $\hat{\psi}$ by plugging in the maximum likelihood estimate $\hat{\psi}$ for ψ

$$\widehat{s.e.}(\hat{\psi}_i)^2 = \left(\sum_{i^*=1}^{k} \nabla \log L_{i^*}(\hat{\psi}) \nabla \log L_{i^*}(\hat{\psi})' \right)^{-1}_i.$$

From here, the estimated standard error for the linear predictor $\hat{\eta}_{ij}$ follows easily as

$$s.e.(\hat{\eta}_{ij}) = \sqrt{z_i' \left(\sum_{i^*=1}^{k} \nabla \log L_{i^*}(\hat{\psi}) \nabla \log L_{i^*}(\hat{\psi})' \right)^{-1}_j z_i}.$$

6.2 Application of the model

As illustration of the model we will use the data of the quitting smoking study (see also Table 4.1). Without covariate information three mixture components can be found in the 59 trials (see Table 6.1). This is in fact the nonparametric profile maximum likelihood estimator, since the associated maximum of the gradient function is bounded by one.

In Table 6.2 we consider the covariate NRT (patch vs. gum) and assume that

Table 6.2 *Does treatment modification change the effect of NRT on quitting smoking? Results of fitting a mixture of two components of regression model on the effect of type of NRT on quitting smoking*

Mixed Component	weight q_j	Covariates	$\hat{\beta}_j$	S.E.	P-value
1	0.8488	Intercept$_1$	0.3357	0.0688	0.0000
		NRT$_1$	0.2253	0.1001	0.0122
2	0.1512	Intercept$_2$	0.6607	0.1660	0.0000
		NRT$_2$	1.0701	0.6632	0.0533

Log-Likelihood $= -17,214.1$
BIC $= -34,444.5$

Table 6.3 *Does treatment modification change the effect of NRT on quitting smoking? Results of fitting a mixture of three components of regression model on the effect of type of NRT on quitting smoking*

Mixed Component	weight q_j	Covariates	$\hat{\beta}_j$	S.E.	P-value
1	0.5244	Intercept$_1$	0.2564	0.1023	0.0061
		NRT$_1$	0.3145	0.1804	0.0406
2	0.0994	Intercept$_2$	0.5701	0.3497	0.0515
		NRT$_2$	1.2734	0.7680	0.0486
3	0.3762	Intercept$_3$	1.8435	0.1567	0.0001
		NRT$_3$	0.0008	0.2468	0.4987

Log-Likelihood $= -17,213.5$
BIC $= -34,451.4$

the mixture of two components exist in this MAIPD of 59 trials. The covariate NRT is only significant in the first component. In the second component the treatment effect is larger than in the first component. If we consider the mixture of three components in Table 6.3 the gradient function is less than one in the parameter space, so that this solution already provides the non-parametric maximum likelihood estimator. A comparison of the BIC values shows that the mixture of two components of regression model provide the preferred statistical model.

The theory of the *general mixture maximum likelihood theorem* is also applicable here but the gradient function is multidimensional. With one covariable, the gradient function has two dimensions.

6.3 Simplification of the model for observed and unobserved covariates

The result in Table 6.3 shows that the effects of the covariate vary from component to component in the mixture model. However, the model will be simple if the effects of covariates are identical over all trials. The idea is to estimate a homogeneous effect of the covariates and allow different intercepts (heterogeneity) for the various mixture components.

Let us rewrite the likelihood function from (6.9) as

$$L(Q(B^*)) = \prod_{i=1}^{k} \sum_{j=1}^{m} f_i(\exp(\eta_{ij}))q_{j\cdot}, \qquad (6.21)$$

with the linear predictor

$$\eta_{ij} = \beta_{0j} + \beta_1 z_{i1} + ... + \beta_p z_{ip}$$

and where $Q(B^*)$ is identical with 6.7 and

$$B^* = \begin{pmatrix} \beta_{01} & \beta_1... & \beta_p \\ ... & ... & ... \\ \beta_{0m} & \beta_1... & \beta_p \end{pmatrix}.$$

Note that $\beta_1, ..., \beta_p$ are the same in each mixture component whereas $\beta_{01}, ..., \beta_{0m}$ are the possible different intercepts in the m mixture components. The model now has $m + p$ regression coefficients and q weights.

6.3.1 Finding the maximum likelihood estimator

The estimation process goes the same way as Section 6.1.1 up to the maximization of the expected complete data log-likelihood (6.13), given as

$$
\begin{aligned}
L(Q(B)) &= \sum_{i=1}^{k} \sum_{j=1}^{m} e_{ij} \log f_i(\exp(\eta_{ij})) + e_{ij} \log(q_j) \\
&= \sum_{i=1}^{k} \sum_{j=1}^{m} e_{ij} \left(x^T \eta_{ij} - (x_i^C + x_i^T) \log(n_i^C + \exp(\eta_{ij})n_i^T)\right) + e_{ij} \log(q_j) \\
&= \sum_{j=1}^{m} \sum_{i=1}^{k} e_{ij} \left(x_i^T \eta_{ij} - (x_i^C + x_i^T) \log(n_i^C + \exp(\eta_{ij})n_i^T)\right) + e_{ij} \log(q_j).
\end{aligned}
$$

$$(6.22)$$

Maximizing of (6.22) with respect to q_j is easily done as previously to provide

$$\hat{q}_j = \frac{1}{k} \sum_{i=1}^{k} e_{ij}.$$

Maximization for

$$\sum_{j=1}^{m}\sum_{i=1}^{k} e_{ij}\left(x_i^T \eta_{ij} - (x_i^C + x_i^T)\log(n_i^C + \exp(\eta_{ij})n_i^T)\right)$$

can be completed in *two* steps.

- *Step 1.* Assume that the common slopes $\beta_1, ..., \beta_p$ are given. Then maximization of (6.22) can be completed via the Newton-Raphson iteration. This can be accomplished by maximizing separately for each component

$$L(\beta_{0j}) = \sum_{i=1}^{k} e_{ij}\left(x_i^T \eta_{ij} - (x_i^C + x_i^T)\log(n_i^C + \exp(\eta_{ij})n_i^T)\right),$$

so that Step 1 is completed by doing m maximizations of $L(\beta_{0j})$ for $j = 1, ..., m$ by means of m univariate Newton-Raphson iteration

$$\beta_{0j}^{(n+1)} = \beta_{0j}^{(n)} - \frac{L'(\beta_{0j}^{(n)})}{L''(\beta_{0j}^{(n)})}$$

which has to be iterated until convergence. Note that these derivatives are easily available as

$$L'(\beta_{0j}) = \sum_{i=1}^{k} e_{ij}\left(x_i^T - (x_i^C + x_i^T)n_i^T \frac{\exp(\eta_{ij})}{n_i^C + \exp(\eta_{ij})n_i^T}\right) \qquad (6.23)$$

and

$$L''(\beta_{0j}) = -\sum_{i=1}^{k} e_{ij}\left(\frac{(x_i^C + x_i^T)n_i^T \exp(\eta_{ij})}{\left(n_i^C + \exp(\eta_{ij})n_i^T\right)^2}\right). \qquad (6.24)$$

- *Step 2.* Now assume that $\beta_{01}, ..., \beta_{0m}$ is given. Then the expected, complete data log-likelihood no longer separates into components, since each component contains the common slopes $\beta_1, ..., \beta_p$:

$$\frac{\partial}{\partial \beta_{d1}}L(\beta_1, ..., \beta_p) = \sum_{i=1}^{k}\sum_{j=1}^{m} e_{ij}\left(x_i^T - (x_i^C + x_i^T)n_i^T \frac{\exp(\eta_{ij})}{n_i^C + \exp(\eta_{ij})n_i^T}\right) z_{i,d1}$$

$$(6.25)$$

$$\frac{\partial^2}{\partial \beta_{d2}\partial \beta_{d1}}L(\beta_1, ..., \beta_p) = -\sum_{i=1}^{k}\sum_{j=1}^{m} e_{ij}\frac{(x_i^C + x_i^T)n_i^T \exp(\eta_{ij})}{\left(n_i^C + \exp(\eta_{ij})n_i^T\right)^2}z_{i,d1}z_{i,d2}.$$

These elements form the gradient

$$\nabla L(\beta) = \left(\frac{\partial}{\partial \beta_{d1}}L(\beta_1, ..., \beta_p)\right)$$

and the Hessian matrix

$$\nabla^2 L(\beta) = \left(\frac{\partial^2}{\partial \beta_{d2}\partial \beta_{d1}}L(\beta_1, ..., \beta_p)\right).$$

Table 6.4 *Does treatment modification change the effect of NRT on quitting smoking?*
Results of fitting a mixture of two components of regression model with a common
slope on the effect of type of NRT on quitting smoking

Mixed Component	weight q_j	Covariates	$\hat{\beta}_j$	S.E.	P-value
1	0.6067	Intercept_1	0.2803	0.3167	0.1881
		NRT	0.2447	0.2873	0.1972
2	0.3933	Intercept_2	0.5815	0.3745	0.0602
		NRT	0.2447	1.3511	0.4281

Log-Likelihood $= -17,214.4$
BIC $= -34,445.1$

so that the Newton-Raphson iteration $\beta^{(n+1)} = \beta^{(n)} - \nabla^2 L(\beta^{(n)})^{-1} \nabla L(\beta^{(n)})$
can be employed. Note also that lower bound methods are possible as in
the section on separate component slopes which leads to

$$\beta^{(n+1)} = \beta^{(n)} + \mathbf{B}^{-1} \nabla L(\beta^{(n)}) \tag{6.26}$$

where $\mathbf{B}$ can be written as $Z'\Lambda Z$, where Λ is a diagonal matrix with $\Lambda_{ii} = \sum_{j=1}^{m} c_{ij} (x_i^T + x_i^C)\frac{1}{4}$, independent of η. $\mathbf{B}$ represents a global upper bound
for $-\nabla^2 L(\beta)$.

Standard errors can be found in a very similar way as outlined in Section 6.1.3.
Let us illustrate the model with the MAIPD on the effect of NRT on quitting
smoking as presented in Table 6.4. Note that the BIC value of $-34,444.5$
is very close to the BIC value of $-34,445.1$ for the two-component model
allowing for different slopes (see Table 6.2). Since a common slope has clear
interpretative advantages, and given the similar empirical support for both
models, the common slope model might be the preferred choice in this case.

CHAPTER 7

Working with CAMAP

The profile log-likelihoods that have been developed and used in this book are nonstandard log-likelihoods. They are neither a Poisson log-likelihood nor any of the log-likelihoods available in the standard generalized linear model family. This makes them less attractive to use one of the existing statistical packages such as STATA, S-plus, MINITAB, or any other package that offers macro-like programming.

The purpose of the present chapter is to provide a very flexible tool for estimating the relative risk based upon the profile log-likelihood models. It was decided to use the Microsoft Fortran Power Station to develop a software tool. This software tool is called CAMAP (Computer Assisted Analysis of Meta-Analysis using the Profile Likelihood Model). The reason for this is the Microsoft Fortran Power Station has many features that make development easy and efficient. The following features have been developed:

- The first feature is to calculate relative risk based upon the basis of the profile log-likelihood model. This feature enables calculation of relative risk based upon the situation of homogeneity effect.

- The second feature is to calculate relative risk based upon the modeling of unobserved heterogeneity. This feature enables calculation of relative risk based upon the mixture of profile log-likelihood model without covariate information.

- The third feature is to calculate relative risk based upon the modeling of covariate information. This feature enables incorporation of covariate information based upon a modification of the generalized linear model in order to estimate the relative risk based upon the significant covariates.

- The fourth feature is to calculate relative risk based upon the modeling of the covariate information to allow for unobserved heterogeneity.

We will give some examples that demonstrate how to work with CAMAP to analyze the data from the meta-analysis of 59 trials that evaluate the effect of nicotine replacement therapy (NRT) on quitting smoking given by DuMouchel and Normand (2000). The data are displayed in Table 4.1. There are two covariates with values defined for each study; the forms of NRT (patch and gum) and the types of support (high support and low support) that might explain the heterogeneity among trials.

7.1 Getting started with CAMAP

7.1.1 Starting

When you execute CAMAP, the following display, shown in Figure 7.1, will appear on your screen. In the header bar at the top of the screen is a list of topics: File, Calculation, and Settings. There are two windows.

- Start window: displays the data set from the input file and shows the output from the analysis.
- Status window: displays the steps of computation.

7.1.2 Importing data

- You can import data into CAMAP in the form of ASCII files saved from a spreadsheet.
- An example of an ASCII file for the meta-analysis to evaluate the effect of NRT on quitting smoking is shown in Figure 7.2. The meaning of each column is:
  - The first column is the number of centers.
  - The second column is the number of smokers who quit smoking in the treatment arm.
  - The third column is the number of smokers in the treatment arm.
  - The fourth column is the number of smokers who quit smoking in the control arm.
  - The fifth column is the number of smokers in the control arm.
  - The sixth column is the binary covariate that describes the forms of NRT.
  - The seventh column is the binary covariate that describes the types of support.
- To import your data file into CAMAP, click on the File menu in the header bar and select the **Open Data** item.
- Having selected a data file from the Input Data File dialog box, a Read Data dialog box will appear with a place for specifying the number of covariates.
- By clicking on the counting check box on the first line, you can tell the program that you record the number of centers in the first column.
- Suppose we opened the previous ASCII file for the meta-analysis to evaluate the effect of NRT on quitting smoking, we filled in the number of covariates as two and clicked on the counting check box. The Read Data dialog box of this example is shown in Figure 7.3 and the data are shown on the Start window screen in Figure 7.4.

7.1.3 Creating a log file for your own work

- It is generally advisable to keep a log of your work session because you can then review what you have done.

- To create a log file in which your outputs are recorded, click on the File menu in the header bar and select the **Start Log-Session** item. A Start Log-Session dialog box will appear with a place for a filename which is shown in Figure 7.5. A filename is any name you wish to give your log file.

- Having finished the computations, click on the File menu in the header bar and select the **Close Log-Session** item. This action will close your log file and all of your outputs will be saved in a text format.

- By clicking on the check box for the Output in LaTeX Format, you can tell the program that you want to record your output in a LaTeX format instead of text format.

- Reviewing the log file can be done with any editor. You can use Wordpad, Notepad, etc., to open and read it.

7.1.4 Exiting CAMAP

When you exit CAMAP, your log file will be automatically saved. You click on the File menu in the header bar and click on the **Exit** item. CAMAP exits by this action.

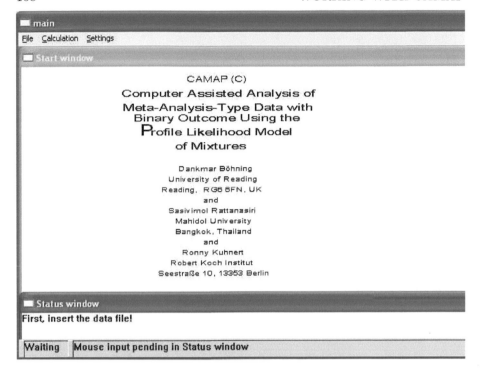

Figure 7.1 *CAMAP window layout*

```
 NRT_cov.txt - Notepad                                    _ □ ✕
File  Edit  Format  View  Help

1      29     116     21     113     0     1
2      6      73      3      121     0     1
3      30     50      23     50      0     1
4      23     180     15     172     0     1
5      22     58      9      58      0     1
6      31     106     16     100     0     1
7      16     44      6      20      0     1
8      9      30      6      30      0     1
9      30     71      14     68      0     1
10     23     60      12     53      0     1
11     37     92      24     90      0     1
12     21     68      5      38      0     1
13     129    600     112    617     0     1
14     13     30      5      30      0     1
15     21     107     21     105     0     1
16     90     211     28     82      0     1
17     51     146     40     127     0     1
18     75     206     50     211     0     1
```

Figure 7.2 *Example of ASCII file of the meta-analysis for evaluating the effect of NRT on quitting smoking*

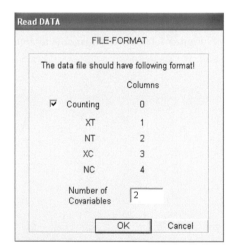

Figure 7.3 *Example of Read Data Dialog Box of the meta-analysis for evaluating the effect of NRT on quitting smoking*

■ main

File Calculation

■ Start window

Datafile: D:\Program\NRT_cov.txt

		Treatment arm		Control arm	Covariables	
Center	xt	nt	xc	nc	cov1	cov2
1	29	116	21	113	0	1
2	6	73	3	121	0	1
3	30	50	23	50	0	1
4	23	180	15	172	0	1
5	22	58	9	58	0	1
6	31	106	16	100	0	1
7	16	44	6	20	0	1
8	9	30	6	30	0	1
9	30	71	14	68	0	1
10	23	60	12	53	0	1
11	37	92	24	90	0	1
12	21	68	5	38	0	1
13	129	600	112	617	0	1
14	13	30	5	30	0	1
15	21	107	21	105	0	1
16	90	211	28	82	0	1
17	51	146	40	127	0	1
18	75	206	50	211	0	1

■ Status window

First, insert the data file!
Data available.

Waiting	Mouse input pending in Status window

Figure 7.4 *Example of output for importing the data from the meta-analysis for evaluating the effect of NRT on quitting smoking*

Figure 7.5 *Start Log-Session Dialog Box*

7.2 Analysis of modeling

- All the features are available from the **Calculation** menu and its associated dialogs.

- To estimate the relative risk based upon any profile likelihood models that are described in this book, click on the **Calculating Relative Risk** item on the Calculation menu.

- A modeling dialog box will appear, giving you a place to enter the stop criteria for the EM algorithm. The default stop criteria is 0.00000000001.

- The criteria for selecting the appropriate number of components; a) NPMLE b) BIC or c) specific maximum number of mixed components, needs to be specified in the modeling dialog box.

  - The NPMLE and BIC are usually used when the number of components is estimated.

  - The specific maximum number of mixed components is usually used when you want to specify the number of components.

- If you have covariates in your data file, the name of the covariates shows in the option–Select included Covariates.

- When the computations are complete, CAMAP displays the output in the Results Window.

- Any profile likelihood models that are described in this book can be executed by CAMAP. We will start with the profile likelihood model for the situation of the homogeneity effect, and then we will demonstrate the more complicated profile likelihood models; the modeling of unobserved heterogeneity, the modeling of covariate information, and the modeling of covariate information to allow for unobserved heterogeneity, respectively.

7.2.1 Modeling of the homogeneity effect

This feature enables calculation of the relative risk based upon the profile likelihood model for the situation of the homogeneity which is described in Chapter 2.

- In the **modeling** dialog box, you click on the radio button for the MAX Number of Mixed Components and enter the maximum number of mixed components as one.

- The completed **modeling** dialog box for the modeling of homogeneity effect of the meta-analysis for evaluating the effect of NRT on quitting smoking is shown in Figure 7.6.

- The results of pressing the OK button with this setup are shown in Figure 7.7.

- The estimate of the effect parameter is equal to 0.4483 with a standard error equal to 0.0392. This corresponds to an estimate of 1.5656 with a 95% confidence interval from 1.4500 to 1.6905 for the relative risk itself.

7.2.2 Modeling of the unobserved heterogeneity

This feature enables calculation of the relative risk based upon the mixture of profile log-likelihood model without covariate information which is described in Chapter 3. First, we try the common approach of a mixture model that uses NPMLE for estimating the number of components.

- The completed **modeling** dialog box to compute the relative risk based upon the modeling of unobserved heterogeneity using the NPMLE of the meta-analysis for evaluating the effect of NRT on quitting smoking is shown in Figure 7.8.
- The results of pressing the OK button with this setup are shown in Figure 7.9.
- The results provide some evidence of heterogeneity consisting of three components.

  The estimate of the effect parameter in the first component is equal to 0.5654 with a standard error equal to 0.0668.
  - The estimate of the effect parameter in the second component is equal to 0.2633 with a standard error equal to 0.0749.
  - The estimate of the effect parameter in the third component is equal to 1.6707 with a standard error equal to 0.7410.

- Having found the mixing distribution and its estimate parameters, the next step is to classify the effect parameter for each trial into one of the components of the mixing distribution which is described in 3.6 of Chapter 3.
- Fifty-two trials are classified into the first component with an estimated relative risk of 1.7601 and a 95% confidence interval from 1.5440 to 2.0065. Seven trials are classified into the second component with an estimated relative risk of 1.3012 and a 95% confidence interval from 1.1234 to 1.5071. However, nontrials are classified into the third component.

However, the result from this example is unsatisfactory, a better result can be obtained by computing the relative risk based upon the modeling of unobserved heterogeneity by specifying the maximum number of mixed components as two.

- The completed modeling dialog box to compute the relative risk based upon the modeling of unobserved heterogeneity by specifying the number of components as two of the meta-analysis for evaluating the effect of NRT on quitting smoking is shown in Figure 7.10.

Figure 7.6 *The setup screen of the modeling of homogeneity effect for the meta-analysis evaluating the effect of NRT on quitting smoking*

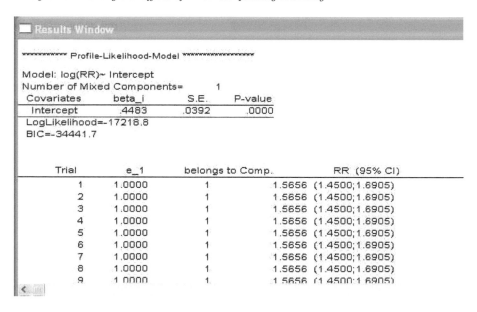

Figure 7.7 *The results screen of the modeling of homogeneity effect for the meta-analysis evaluating the effect of NRT on quitting smoking*

Figure 7.8 *The setup screen of the modeling of unobserved heterogeneity for the meta-analysis evaluating the effect of NRT on quitting smoking*

Results Window

~~~~~~~~~~~ Profile-Likelihood-Model ~~~~~~~~~~~~~~~~~~~

Model: log(RR)~ Intercept
Number of Mixed Components=　　　3

Mixed Comp.	Weights	S.E. of W	N(Trials)	Covariates	beta_i	S.E.	P-value
1	.6418	.2825	52	Intercept	.5654	.1001	.0000
2	.3342	.2816	7	Intercept	.2633	.1182	.0129
3	.0240		0	Intercept	1.6707	.7794	.0160

LogLikelihood=-17216.9
BIC=-34454.1
Max of Gradient function=1.0000

Trial	e_1	e_2	e_3	belongs to Comp.	RR (95% CI)
1	.5564	.4436	.0000	1	1.7601 (1.4464;2.1418)
2	.7366	.2293	.0341	1	1.7601 (1.4464;2.1418)
3	.5194	.4806	.0000	1	1.7601 (1.4464;2.1418)
4	.6379	.3621	.0000	1	1.7601 (1.4464;2.1418)
5	.8336	.1583	.0081	1	1.7601 (1.4464;2.1418)
6	.7813	.2185	.0002	1	1.7601 (1.4464;2.1418)

Figure 7.9 *The results screen of the modeling of unobserved heterogeneity for the meta-analysis evaluating the effect of NRT on quitting smoking*

- The results of pressing the OK button with this setup are shown in Figure 7.11.

  - The estimate of the effect parameter in the first component is equal to 0.5654 with a standard error equal to 0.0668.
  - The estimate of the effect parameter in the second component is equal to 0.2633 with a standard error equal to 0.0749.

- Forty-nine trials are classified into the first component with an estimated relative risk of 1.8010 and a 95% confidence interval from 1.4900 to 2.1771. Ten trials are classified into the second component with an estimated relative risk of 1.3030 and a 95% confidence interval from 1.0406 to 1.6315.

7.2.3 Modeling of the covariate information

This feature enables incorporation of covariate information based upon a modification of the generalized linear model using the profile likelihood approach in order to estimate the relative risk based upon the significant covariates that is described in Chapter 4.

- In the **modeling** dialog box, you click on the radio button for the MAX Number of Mixed Components and enter the maximum number of mixed components as one in this case.
- You will then need to select covariates that you want to consider from the option– Select included covariate.
- The completed modeling dialog box for computing the relative risk based upon the forms of NRT and the types of support using the profile likelihood approach is shown in Figure 7.12.
- The results of pressing the OK button with this setup are shown in Figure 7.13.
- The estimate of the effect parameter is equal to 0.2633 with a standard error equal to 0.0749 for the forms of NRT, and the estimate of the effect parameter is equal to 0.2633 with a standard error equal to 0.0749 for the types of support.
- These correspond to an estimated relative risk as follows:

  - An estimated relative risk of 1.4138 with a 95% confidence interval from 1.2608 to 1.5854 for gum and high support groups.
  - An estimated relative risk of 1.8201 with a 95% confidence interval from 1.5640 to 2.1181 for patch and high support groups.
  - An estimated relative risk of 1.5458 with a 95% confidence interval from 1.3581 to 1.7596 for gum and low support groups.
  - An estimated relative risk of 1.9900 with a 95% confidence interval from 1.6282 to 2.4323 for patch and low support groups.

Figure 7.10 *The setup screen of the modeling of unobserved heterogeneity for the meta-analysis evaluating the effect of NRT on quitting smoking*

```
■ Results Window

wwwwwwww Profile-Likelihood-Model wwwwwwwwwwwwwwww

Model: log(RR)~ Intercept
Number of Mixed Components=        2
Mixed Comp.    Weights   S.E. of W    N(Trials)   Covariates     beta_i      S.E.     P-value
        1       .6525      .2740           49     Intercept       .5884      .0967      .0000
        2       .3475      .2740           10     Intercept       .2646      .1147      .0105
LogLikelihood=-17217.0
BIC=-34446.3
Max of Gradient function=1.3030


      Trial         e_1         e_2      belongs to Comp.          RR  (95% CI)
        1         .5318       .4682            1           1.8010  (1.4900;2.1771)
        2         .7637       .2363            1           1.8010  (1.4900;2.1771)
        3         .4912       .5088            2           1.3030  (1.0406;1.6315)
        4         .6231       .3769            1           1.8010  (1.4900;2.1771)
        5         .8432       .1568            1           1.8010  (1.4900;2.1771)
        6         .7778       .2222            1           1.8010  (1.4900;2.1771)
        7         5794        4206            1           1 8010  (1 4900·2 1771)
```

Figure 7.11 *The results screen of the modeling of unobserved heterogeneity for the meta-analysis evaluating the effect of NRT on quitting smoking*

- However, the results indicate that the forms of NRT yield the only significant change for the treatment effect of quitting smoking. A better result can be obtained from estimating the relative risk based only upon the forms of NRT that are shown in Table 4.3.

7.2.4 Modeling of the covariate information to allow for unobserved heterogeneity

This feature enables calculation of relative risk based upon the modeling of covariate information to allow for unobserved heterogeneity which is described in Chapter 6. First, we consider the more general situation that both effects of treatment and covariates vary from component to component in the mixture model.

- In the **modeling** dialog box, you have to specify the criteria for selecting the appropriate number of components.
- You will then need to select covariates that you want to consider from the option–Select included covariate.
- The completed modeling dialog box to compute the relative risk based upon modeling of the forms of NRT to allow for unobserved heterogeneity by using NPMLE is shown in Figure 7.14.
- The results of pressing the OK button with this setup are shown in Figure 7.15.

Alternatively, the model will be simple if the effects of covariates are identical over all trials. The idea is to estimate the relative risk based upon a homogeneity effect of the covariates and allow for the heterogeneity effect of the treatment which is described in 6.3 of Chapter 6.

- In the **modeling** dialog box, you have to specify the criteria for selecting the appropriate number of components and select covariates that you want to consider as in the previous model.
- You will then need to click on the simplify model check box.
- The completed modeling dialog box to compute the relative risk based upon the simple modeling of the forms of NRT to allow for unobserved heterogeneity by using NPMLE is shown in Figure 7.16.
- The results of pressing the OK button with this setup are shown in Figure 7.17.

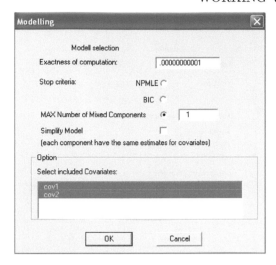

Figure 7.12 *The setup screen of the modeling of covariate information for the meta-analysis evaluating the effect of NRT on quitting smoking*

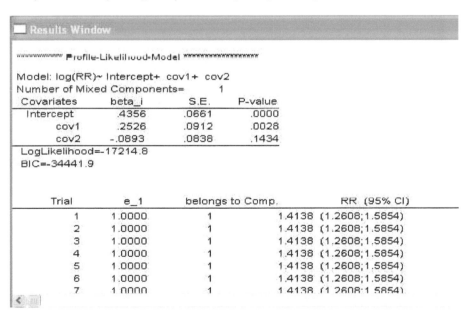

Figure 7.13 *The results screen of the modeling of covariate information for the meta-analysis evaluating the effect of NRT on quitting smoking*

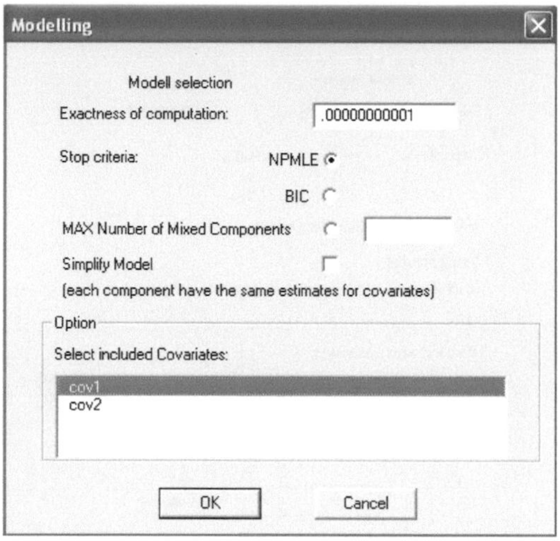

Modelling ☒

Modell selection

Exactness of computation: `.00000000001`

Stop criteria: NPMLE ⦿

BIC ○

MAX Number of Mixed Components ○ [   ]

Simplify Model ☐
(each component have the same estimates for covariates)

Option

Select included Covariates:

cov1
cov2

[OK]   [Cancel]

Figure 7.14 *The setup screen of the modeling of covariate information to allow for unobserved heterogeneity of the meta-analysis for evaluating the effect of NRT on quitting smoking*

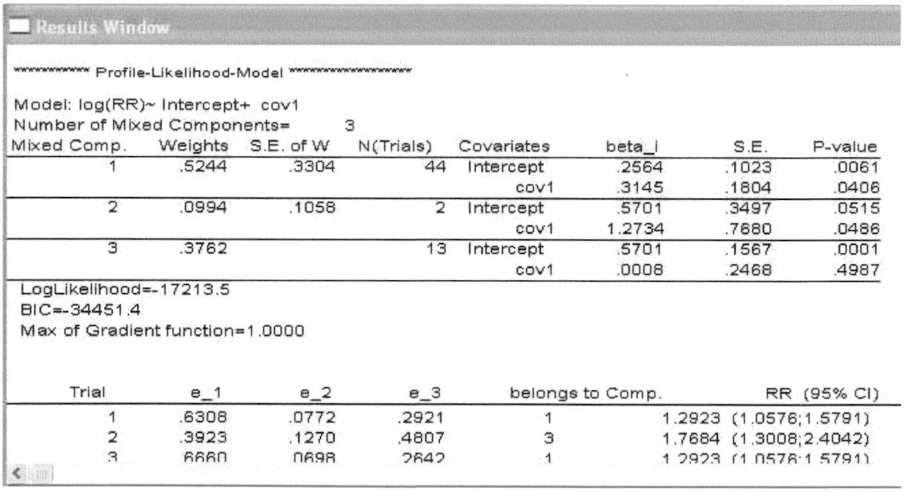

Results Window

*********** Profile-Likelihood-Model ****************

Model: log(RR)~ Intercept+ cov1
Number of Mixed Components= 3

Mixed Comp.	Weights	S.E. of W	N(Trials)	Covariates	beta_i	S.E.	P-value
1	.5244	.3304	44	Intercept	.2564	.1023	.0061
				cov1	.3145	.1804	.0406
2	.0994	.1058	2	Intercept	.5701	.3497	.0515
				cov1	1.2734	.7680	.0486
3	.3762		13	Intercept	.5701	.1567	.0001
				cov1	.0008	.2468	.4987

LogLikelihood=-17213.5
BIC=-34451.4
Max of Gradient function=1.0000

Trial	e_1	e_2	e_3	belongs to Comp.	RR (95% CI)
1	.6308	.0772	.2921	1	1.2923 (1.0576;1.5791)
2	.3923	.1270	.4807	3	1.7684 (1.3008;2.4042)
3	.6660	.0698	.2642	1	1.2923 (1.0576;1.5791)

Figure 7.15 *The results screen of the modeling of covariate information to allow for unobserved heterogeneity of the meta-analysis for evaluating the effect of NRT on quitting smoking*

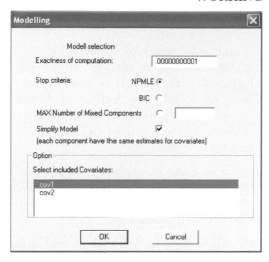

Figure 7.16 *The setup screen of the simple modeling of covariate information to allow for unobserved heterogeneity of the meta-analysis for evaluating the effect of NRT on quitting smoking*

■ **Results Window**

************ Profile-Likelihood-Model ******************

Simplified Procedure
Model: log(RR)~ Intercept+ cov1
Number of Mixed Components=   3

Mixed Comp.	Weights	S.E. of W	N(Trials)	Covariates	beta_i	S.E.	P-value
1	.6052	.9555	42	Intercept	.2807	.2220	.1031
				cov1	.2434	.2108	.1242
2	.3915	.9703	17	Intercept	.5774	.2518	.0109
				cov1	.2434	.8717	.3901
3	.0033		0	Intercept	1.4084	2.2745	.2679
				cov1	.2434	2.1038	.4540

LogLikelihood=-17214.4
BIC=-34453.2
Max of Gradient function=1.0000

Trial	e_1	e_2	e_3	belongs to Comp.	RR (95% CI)
1	.7121	.2879	.0000	1	1.3240 (.8569;2.0458)
2	.4820	.5119	.0061	2	1.7814 (1.0876;2.9180)

Figure 7.17 *The results screen of the simple modeling of covariate information to allow for unobserved heterogeneity of the meta-analysis for evaluating the effect of NRT on quitting smoking*

7.3 Conclusion

The meta-analysis and meta-regression analysis have been demonstrated to provide a powerful statistical tool to analyze and potentially combine the results from individual studies. However, they have some deficiencies and disadvantages in combining the results from individual studies that we mentioned in the previous chapters. The profile likelihood approach as an alternative approach overcomes some of these deficiencies and provides more reliable information on an intervention effect.

However, computation by using this approach is not available in the existing statistical packages. The software tool provided here should allow researchers to accommodate all of the profile likelihood approaches that we explained in this book in a straightforward manner.

The software CAMAP can be downloaded with no costs from:

www.reading.ac.uk/sns05dab/software.html

CHAPTER 8

Estimation of odds ratio using the profile likelihood

Besides the *relative risk* the *Odds Ratio* (OR) is frequently used in epidemiology as another measure of effect. The OR has been mentioned already in Section 1.6.3. The profile likelihood framework can be developed very much the same way as in the previous chapters. The OR is defined as

$$\kappa_i = \frac{q_i^T}{q_i^C},$$

where $q_i^T = p_i^T/(1-p_i^T)$ and $q_i^C = p_i^C/(1-p_i^C)$. For the framework of the odds ratio it is assumed that the observations are binomially distributed. Therefore, the log binomial likelihood is given as (ignoring the only data-dependent term)

$$
\begin{aligned}
ll(p_i^T, p_i^C) = {} & x_i^T \log(p_i^T) + (n_i^T - x_i^T) \log(1 - p_i^T) \\
& + x_i^C \log(p_i^C) + (n_i^C - x_i^C) \log(1 - p_i^C).
\end{aligned}
\tag{8.1}
$$

In the first step p_i^T and p_i^C are replaced by q_i^T and q_i^C. From the condition $q_i^T = \frac{p_i^T}{1-p_i^T}$ it follows that $p_i^T = \frac{q_i^T}{1+q_i^T}$ (for q_i^C similarly) so that the log-likelihood function becomes

$$
\begin{aligned}
ll(q_i^T, q_i^C) = {} & x_i^T \log(\frac{q_i^T}{1 + q_i^T}) + (n_i^T - x_i^T) \log(1 - \frac{q_i^T}{1 + q_i^T}) \\
& + x_i^C \log(\frac{q_i^C}{1 + q_i^C}) + (n_i^C - x_i^C) \log(1 - \frac{q_i^C}{1 + q_i^C}) \\
= {} & x_i^T \log(q_i^T) - n_i^T \log(1 + q_i^T) + x_i^C \log(q_i^C) - n_i^C \log(1 + q_i^C).
\end{aligned}
\tag{8.2}
$$

In the next step, the parameter of interest κ_i is inserted in (8.2) for q_i^T as $\kappa_i q_i^C$ leading to

$$
ll(q_i^T, \kappa_i) = x_i^T \log(\kappa_i) - n_i^T \log(1 + \kappa_i q_i^C) + (x_i^T + x_i^C) \log(q_i^C) - n_i^C \log(1 + q_i^C).
\tag{8.3}
$$

Now it is possible to eliminate the nuisance parameter q_i^C by maximizing (8.3).

The partial derivative of (8.3) with respect to q_i^C is given as

$$\frac{\partial}{\partial q_i^C} ll(q_i^T, \kappa_i) = \frac{x_i^T + x_i^C}{q_i^C} - \frac{n_i^C}{1 + q_i^C} - \frac{n_i^T \kappa_i}{1 + \kappa_i q_i^C}.$$

If the partial derivative of (8.3) is set to zero and solved for q_i^C, two solutions are found as

$$q_i^C[\kappa_i]_1 = -\frac{2(x_i^T + x_i^C)}{x_i^T + x_i^C - n_i^C + (x_i^T + x_i^C - n_i^T)\kappa_i - \sqrt{r_i(\kappa_i)}} \qquad (8.4)$$

$$q_i^C[\kappa_i]_2 = -\frac{2(x_i^T + x_i^C)}{x_i^T + x_i^C - n_i^C + (x_i^T + x_i^C - n_i^T)\kappa_i + \sqrt{r_i(\kappa_i)}} \qquad (8.5)$$

$$r_i(\kappa) = -4(x_i^T + x_i^C)(x_i^T + x_i^C - n_i^T - n_i^C)\kappa$$
$$+ (x_i^T + x_i^C - n_i^C + (x_i^T + x_i^C - n_i^T)\kappa)^2. \qquad (8.6)$$

The second solution $q_i^C[\kappa_i]_2$ is negative for every κ_i, for example $q_i^C[1]_2 = -1$. Only $q_i^C[\kappa_i]_1 := q_i^C[\kappa_i]$ provides a feasible solution in this situation. The resulting profile likelihood function takes the form

$$ll_{PL}(\kappa_i) = x_i^T \log(\kappa_i) - n_i^T \log(1 + \kappa_i q_i^C[\kappa_i])$$
$$+ (x_i^T + x_i^C) \log(q_i^C[\kappa_i]) - n_i^C \log(1 + q_i^C[\kappa_i]). \qquad (8.7)$$

To avoid mathematical problems $q_i^C[\kappa_i]$ has to be positive. If we consider the fraction of $q_i^C[\kappa_i]$ it is easy to see that this situation occurs if $x_i^T = 0$ and $x_i^C = 0$. Although this scenario is rare in medical studies, it can occur in MAIPDs with high sparsity data. In the CALGB study (Table 1.8), this case appears in the 20-th center. The other setting that can cause mathematical problems might occur if the denominator of $q_i^C[\kappa_i]$ equals zero. This occurs if $x_i^T = n_i^T$ and $x_i^C = n_i^C$. Consequently, in both cases we add 0.5 to each cell of the associated 2×2 table.

8.1 Profile likelihood under effect homogeneity

The assumption of a common, fixed effect for all centers leads to the profile likelihood function as

$$ll_{PL}(\kappa) = \sum_{i=1}^{k} x_i^T \log(\kappa) - n_i^T \log(1 + \kappa \times q_i^C[\kappa])$$
$$+ (x_i^T + x_i^C) \log(q_i^C[\kappa]) - n_i^C \log(1 + q_i^C[\kappa]). \qquad (8.8)$$

The maximum of (8.8) coincides with the desired profile maximum likelihood estimator (PMLE). Just like in the Poisson profile likelihood, the first derivative of (8.8) equated to zero provides an approximate fixed-point iteration for the PMLE. The first derivate of (8.8) is

$$ll'_{PL}(\kappa) = \sum_{i=1}^{k} \frac{x_i^T}{\kappa} - \frac{n_i^T \left(\kappa \times \left(q_i^C \right)' [\kappa] + q_i^C [\kappa] \right)}{1 + \kappa \times q_i^C [\kappa]}$$

$$+ \frac{\left(x_i^T + x_i^C \right) \left(q_i^C \right)' [\kappa]}{q_i^C [\kappa]} - \frac{n_i^C \left(q_i^C \right)' [\kappa]}{1 + q_i^C [\kappa]} \tag{8.9}$$

with associated fixed-point iteration

$$\kappa = \frac{\sum_{i=1}^{k} x_i^T}{\sum_{i=1}^{k} \frac{n_i^T \left(\kappa \times \left(q_i^C \right)' [\kappa] + q_i^C [\kappa] \right)}{1 + \kappa \times q_i^C [\kappa]} + \frac{n_i^C \left(q_i^C \right)' [\kappa]}{1 + q_i^C [\kappa]} - \frac{\left(x_i^T + x_i^C \right) \left(q_i^C \right)' [\kappa]}{q_i^C [\kappa]}} := \Phi_{OR}(\kappa). \tag{8.10}$$

From (8.9) we can derive an iteration with convergence rate considerably faster than (8.10). Adding the term

$$\sum_{i=1}^{k} \frac{(n_i^C - x_i^C) x_i^T - x_i^T \sqrt{r_i(\kappa)}}{\kappa \sqrt{r_i(\kappa)}}$$

to both sides of the score equation, e.g., (8.9) set to zero, we find that

$$\kappa = \frac{\sum_{i=1}^{k} \frac{(n_i^C - x_i^C) x_i^T}{\sqrt{r_i(\kappa)}}}{\sum_{i=1}^{k} t_i(\kappa)} := \Gamma_{OR}(\kappa) \tag{8.11}$$

$$\text{where } t_i(\kappa) = \frac{n_i^T \left(\kappa \times \left(q_i^C \right)' [\kappa] + q_i^C [\kappa] \right)}{1 + \kappa \times q_i^C [\kappa]} + \frac{n_i^C \left(q_i^C \right)' [\kappa]}{1 + q_i^C [\kappa]}$$

$$- \frac{\left(x_i^T + x_i^C \right) \left(q_i^C \right)' [\kappa]}{q_i^C [\kappa]} + \frac{(n_i^C - x_i^C) x_i^T - x_i^T \sqrt{r_i(\kappa)}}{\kappa \sqrt{r_i(\kappa)}}.$$

We show in the appendix A.3 the interesting fact that (8.11), if considered as an algorithmic construction rule, delivers in the first iteration step $\Gamma(1)$ the common Mantel-Haenszel estimator, namely

$$\hat{\kappa}_{MH} = \frac{\sum_{i=1}^{k} \frac{(n_i^C - x_i^C) x_i^T}{n_i^T + n_i^C}}{\sum_{i=1}^{k} \frac{(n_i^T - x_i^T) x_i^C}{n_i^T + n_i^C}}.$$

As illustration both fixed point iterations are compared in Figure 8.1. $\Gamma_{OR}(\kappa)$ is clearly faster than $\Phi_{OR}(\kappa)$. To get an accuracy of $\varepsilon = 10^{-8}$ four steps are necessary with $\Gamma_{OR}(\kappa)$ and 26 steps are necessary with $\Phi_{OR}(\kappa)$ using an initial value of $\kappa = 1$. For the sake of brevity we do not address the question of convergence and uniqueness for both iteration rules at this stage.

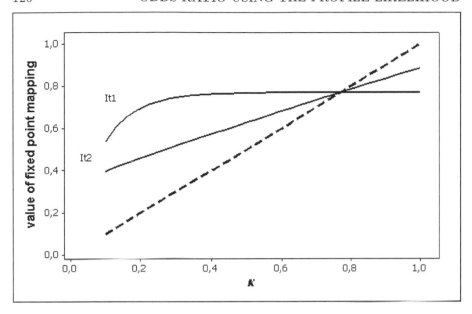

Figure 8.1 *Graph of* $\Gamma_{OR}(\kappa)$ *(It1) and graph of* $\Phi_{OR}(\kappa)$ *(It2) for the data of Table 1.5*

8.2 Modeling covariate information

Analogously to Chapter 4.2 we are interested in the modeling covariate information and use for this purpose the linear predictor $\eta_i = \beta_0 + \beta_1 z_{i1} + ... + \beta_p z_{ip}$, see also 4.18. Using the canonical link $\kappa_i = \exp(\eta_i)$ leads to the profile log-likelihood function

$$ll_{cov}(\eta_i) = ll_{PL}(\exp(\eta_i)) = \sum_{i=1}^{k} x_i^T \eta_i - n_i^T \log(1 + \exp(\eta_i) \times p_i^C[\exp(\eta_i)])$$
$$+ (x_i^T + x_i^C) \log(p_i^C[\exp(\eta_i)]) - n_i^C \log(1 + p_i^C[\exp(\eta_i)]). \qquad (8.12)$$

Here β_0 corresponds to the log(OR) without any covariates in the model. For the estimation it is necessary to maximize (8.12) in $\beta_0, ..., \beta_p$. This can be accomplished using the Newton-Raphson iteration.

8.2.1 The profile likelihood function with covariate information

The derivatives are more complex here, but closed form expressions are possible. The first partial derivative of (8.12) is given as

$$\nabla L(\beta_j) = \frac{\partial}{\partial \beta_j} ll_{cov}(\eta_i) = \sum_{i=1}^{k} x_i^T z_{ij} +$$

$$z_{ij} \exp(\eta_i) \left(-n_i^T t_i^1 - n_i^C t_i^2 + (x_i^T + x_i^C) t_i^3 \right) \qquad (8.13)$$

where

$$t_i^1 = \frac{(\exp(\eta_i) \times (q_i^C)' [\exp(\eta_i)] + q_i^C [\exp(\eta_i)])}{1 + \exp(\eta_i) \times q_i^C [\exp(\eta_i)]}$$

$$t_i^2 = \frac{(q_i^C)' [\exp(\eta_i)]}{1 + q_i^C [\exp(\eta_i)]}$$

$$t_i^3 = \frac{(q_i^C)' [\exp(\eta_i)]}{q_i^C [\exp(\eta_i)]}$$

and the corresponding gradient is

$$\nabla L(\beta) = \left(\frac{\partial}{\partial \beta_0} ll_{cov}(\eta_i), ..., \frac{\partial}{\partial \beta_p} ll_{cov}(\eta_i) \right)^T . \qquad (8.14)$$

The *Hesse* matrix is derived from (8.14) as

$$\nabla^2 L(\beta)_{ih} = \frac{\partial}{\partial \beta_j \beta_h} ll_{cov}(\eta_i) = \sum_{i=1}^{k} z_{ij} z_{ih} \exp(\eta_i)$$

$$\left(-n_i^T \left(t_i^1 - t_i^{1*} \right) - n_i^C \left(t_i^2 - t_i^{2*} \right) + (x_i^T + x_i^C) \left(t_i^3 - t_i^{3*} \right) \right)$$

where

$$t_i^{1*} = \frac{\exp(\eta_i) \times (q_i^C)'' [\exp(\eta_i)] + 2 (q_i^C)' [\exp(\eta_i)] - q_i^C [\exp(\eta_i)]^2}{\left(1 + \exp(\eta_i) \times q_i^C [\exp(\eta_i)] \right)^2}$$

$$+ \frac{\exp(2\eta_i)(- (q_i^C)' [\exp(\eta_i)]^2 + q_i^C [\exp(\eta_i)] (q_i^C)'' [\exp(\eta_i)])}{\left(1 + \exp(\eta_i) \times q_i^C [\exp(\eta_i)] \right)^2}$$

$$t_i^{2*} = \frac{- (q_i^C)' [\exp(\eta_i)]^2 + (1 + q_i^C [\exp(\eta_i)]) (q_i^C)'' [\exp(\eta_i)]}{\left(1 + q_i^C [\exp(\eta_i)] \right)^2}$$

$$t_i^{3*} = \frac{- (q_i^C)' [\exp(\eta_i)]^2 + q_i^C [\exp(\eta_i)] (q_i^C)'' [\exp(\eta_i)]}{q_i^C [\exp(\eta_i)]^2}$$

and as matrix form

$$\nabla^2 L(\beta) = Z^T W Z$$

with

$$
Z = \begin{pmatrix} z_{10} & \cdots & z_{1p} \\ \cdots & \cdots & \cdots \\ z_{k0} & \cdots & z_{kp} \end{pmatrix}
$$

$$
W = \begin{pmatrix} w_1 & \cdots & 0 & \cdots & 0 \\ 0 & \cdots & w_i & \cdots & 0 \\ 0 & \cdots & 0 & \cdots & w_k \end{pmatrix}
$$

where $w_i = \exp(\eta_i) \left(-n_i^T \left(t_i^1 - t_i^{1*} \right) - n_i^C \left(t_i^2 - t_i^{2*} \right) + \left(x_i^T + x_i^C \right) \left(t_i^3 - t_i^{3*} \right) \right).$
The iterative construction of the maximum likelihood estimates for β follows
the iteration

$$
\beta^{(n+1)} = \beta^{(n)} - \nabla^2 L(\beta^{(n)})^{-1} \nabla L(\beta) \tag{8.15}
$$

As a byproduct of the Newton-Raphson iteration (8.15), estimates of the variance of $\hat{\beta}_i$ are available from the i−th diagonal element of the covariance matrix

$$
var(\hat{\beta}_i) = - \left(\nabla^2 L(\hat{\beta}) \right)^{-1}_{ii}
$$

and an associated 95% confidence interval for β_i is achieved from

$$
\exp\{\hat{\beta}_i \pm 1.96 \times \sqrt{var(\hat{\beta}_i)}\}.
$$

8.2.2 Quitting smoking example

As an application of the profile likelihood method for estimating the Odds
Ratio we are using the data of the quitting smoking meta-analysis in Table
4.1.

Table 8.1 *Results of fitting various models to the meta-analysis of quitting smoking*

$L^*(\hat{\beta})$	Covariates	$\hat{\beta}_j$	S.E.	P-value
-7464.64	Intercept	0.5679	0.0439	0.
-7462.13	Intercept	0.50399	0.0523	0.
	NRT	-0.3717	0.2167	0.0128
-7464.64	Intercept	0.5699	0.0711	0.
	Support	-0.0033	0.0904	0.4856
-7461.91	Intercept	0.5373	0.0726	0.
	NRT	0.2345	0.1007	0.0099
	Support	-0.0624	0.0940	0.2536

The results of Table 8.1 are comparable with the results of Table 4.2. NRT is the one covariate with a significant influence on the treatment effect. The variance of the parameters seems to be a little bit higher than the parameters in Table 4.2. Secondly, the P-values are slightly higher in the odds ratio estimation process. Finally, both models are coincided in the model selection and in the treatment predication.

Table 8.2 *Results of estimating the odds ratio with 95% CI in the meta-analysis of quitting smoking*

	S.E.	OR (95% CI)
No covariates	0.0439	1.7645 (1.6191, 1.9230)
Form of NRT		
Patch	0.0818	2.0558 (1.7512, 2.4134)
Gum	0.0523	1.6552 (1.4940, 1.8340)

CHAPTER 9

Quantification of heterogeneity in a MAIPD

9.1 The problem

In many systematic reviews and meta-analyses statistical heterogeneity occurs. This means that the effect measure differs between trials more than it could be expected under homogeneity of effect. However, the question might be not if there is heterogeneity but more how its amount could be quantified. As Higgins and Thompson (2002) put it:

> Addressing statistical heterogeneity is one of the most troublesome aspects of many systematic reviews. The interpretative problems depend on how substantial the heterogeneity is, since this determines the extent to which it might influence the conclusions of the meta-analysis. It is therefore important to be able to quantify the extent of heterogeneity among a collection of studies.

A small amount of heterogeneity in a MAIPD might lead to a valid analysis similar to an analysis under complete homogeneity, whereas a large amount of heterogeneity might require larger efforts in coping with it including a more substantial way of modeling it, as has been suggested in previous chapters. One historic form of quantifying heterogeneity is the Q-statistic usually attributed to Cochran (1954) (see also Whitehead and Whitehead (1991) or Normand (1999)). It is defined in our setting as

$$Q = \sum_{i=1}^{k} w_i(\hat{\phi}_i - \bar{\phi})^2 \qquad (9.1)$$

where $w_i = 1/Var(\hat{\phi}_i)$ for $\phi_i = \log \theta_i$ and

$$\bar{\phi} = \frac{\sum_{i=1}^{k} w_i \hat{\phi}}{\sum_{i=1}^{k} w_i}.$$

The statistic Q is also the basis for estimating the between-studies variance $\tau_\phi^2 = Var(\phi)$ with the moment estimator suggested by DerSimonian and Laird (1986)

$$\tau_{DL}^2 = \frac{Q - (k-1)}{\sum_i w_i - \frac{\sum_i w_i^2}{\sum_i w_i}}$$

if $Q > (k-1)$, and 0 otherwise.

The statistic Q has a χ^2-distribution with $(k-1)$ degree of freedom if $\hat{\phi}_i$ is normally distributed with mean ϕ and variance σ_i^2. However, in our situation some problems arise in using Q as a test statistic:

- The variance of $\hat{\phi}_i$ is usually unknown and is estimated as $\hat{\sigma}_i^2 = 1/x_i^T + 1/x_i^C$.
- The number of events in both trial arms is more realistically assumed to be Poisson rather than normal, so that large number of events are required to have a reasonable approximation to the normal.
- For high sparsity trials zero events are likely to occur and will require some modification of the variance estimate, often achieved by adding 0.5 in both arms.

For a more general discussion of homogeneity tests see Hartung et al. (2003) and for alternative test procedures Hartung and Knapp (2003).

Two issues need to be distinguished:

- First, the asymptotic χ^2-result occurs if the number of trials k is kept fixed and the trial sizes n_i become large.
- Second, if the individual trial size remains small while k increases, Q will only have a χ^2-distribution with $k-1$ df under special circumstances such as characterized by Potthoff and Whittinghill (1966).

Higgins and Thompson (2002) suggest to consider Q as a measure of heterogeneity and define two measures on the basis of Q:

$$H^2 = \frac{Q}{k-1}$$

and

$$I^2 = \frac{H^2 - 1}{H^2}.$$

Leaving inferential issues aside for the time being, several aspects of these measures should be noted:

- H^2 takes values from 0 to ∞ and reflects that Q is not bounded above. It is therefore an absolute measure. Higgins and Thompson (2002) suggest to truncate values of H^2 smaller than 1 at 1, since the expected value of a χ^2 with $k-1$ df is $k-1$.
- This will then make I^2 nonnegative and also bound above by 1. Hence, I^2 is a relative measure. However, this bound will never be attained in practice, since it would require limiting values for Q.

It can be expected that any of these measures, Q, H^2, or I^2, are rather unstable when used in a high sparsity trial as in the following example.

9.1.1 A sparse multicenter hypertension trial

This example is taken from Brown and Prescott (1999). The trial was a randomized double blind comparison of three treatments for hypertension and has been reported in Hall et al. (1991). One treatment was a new drug (A = Carvedilol) and the other two (B = Nifedipine and C = Atenolol) were standard drugs for controlling hypertension. A number of centers participated in the trial and patients were randomized in order of entry. We concentrate here on the adverse event of developing the symptom of *cold feet* during the treatment (for details see Brown and Prescott (1999), p. 137). The data are provided in Table 9.1.

Table 9.1 *Data illustration of a MAIPD for studying the effect of developing cold feet during treatment for hypertension (Hall et al. (1991)); data contain number of patients* x_i^T, x_i^{C1}, x_i^{C2} *developing cold feet and number of patients recruited* n_i^T, n_i^{C1}, n_i^{C2} *in the treatment (T) and control arms (C1 and C2), respectively*

Center	Treatment		Control 1		Control 2	
i	x_i^T	n_i^T	x_i^{C1}	n_i^{C1}	x_i^{C2}	n_i^{C2}
1	3	13	5	14	1	12
2	2	3	0	4	0	3
3	0	3	0	3	0	2
4	1	4	1	4	0	4
5	1	4	3	5	0	2
6	0	2	1	1	1	2
7	0	6	1	6	0	6
8	1	2	0	1	1	2
9	0	4	1	4	0	4
10	0	3	1	3	0	4
11	1	1	0	1	0	2
12	0	8	2	8	1	8
13	1	4	0	4	0	3
14	0	2	0	2	0	2
15	0	3	0	2	0	2
16	0	3	1	4	0	3
17	0	1	0	2	0	2
18	0	12	0	12	0	12
19	1	2	0	1	0	1
20	0	9	5	6	0	8
21	0	2	0	1	1	2
22	0	2	0	1	0	1

In a MAIPD with high sparsity such as in Table 9.1 the application of Q or any other measure based upon Q would be problematic. The estimated weights $\hat{w}_i = \frac{x_i^T x_i^C}{x_i^T + x_i^C}$ would be not defined in many cases where $x_i^T = 0$ or $x_i^C = 0$ and the required modification $\hat{w}_i = \frac{(x_i^T + 0.5)(x_i^C + 0.5)}{x_i^T + x_i^C + 1}$ would leave any measure involving these highly unstable. In the following we develop an alternative approach for measuring heterogeneity in a MAIPD.

9.2 The profile likelihood as binomial likelihood

In Chapter 2 we have developed the profile log-likelihood (2.11) for study i

$$L^*(\theta) = \sum_{i=1}^{k} \{x_i^T \log(\theta_i) - (x_i^C + x_i^T) \log(n_i^C + \theta_i n_i^T)\}$$

which corresponds to a profile likelihood

$$\frac{(n^T \theta)^{x^T}}{(n^C + \theta n^T)^{x^T + x^C}}, \tag{9.2}$$

where we have ignored the index for the study. It is quite interesting to observe that (9.2) is proportional to a *binomial likelihood*

$$\binom{N}{y} p^y (1-p)^{N-y}, \tag{9.3}$$

with $p = \frac{\theta n^T}{n^C + \theta n^T}$ and $N = x^T + x^C$, $y = x^T$. Note that when the trial is balanced $(n^T = n^C = n)$, p and θ are related simply by $p = \theta/(1+\theta)$.

9.3 The unconditional variance and its estimation

It is appropriate to recall a few facts about the binomial distribution with density $f(y|p) = \binom{N}{y} p^y (1-p)^{(N-y)}$. The binomial has mean $E(Y|p) = Np$ and variance $Var(Y|p) = Np(1-p)$. Note that these moments are conditional on the value of p. Suppose now that p itself is random with density $f(p)$ so that the joint and marginal density of Y are given as

$$f(y,p) = \quad f(y|p)f(p) \tag{9.4}$$
$$f(y) = \quad \int_0^1 f(y|p)f(p)dp, \tag{9.5}$$

respectively. We will also denote the mean and variance of p with

$$E(p) = \int_0^1 p f(p) dp = \mu \text{ and } Var(p) = \int_0^1 (p-\mu)^2 f(p) dp = \tau^2.$$

It follows that the marginal mean and variance of Y are provided as

$$E(Y) = \qquad E[E(Y|p)] = N\mu \qquad (9.6)$$
$$Var(Y) = \quad E[Var(Y|p)] + Var[E(Y|p)]. \qquad (9.7)$$

Equation (9.7) can be further developed to become

$$Var(Y) = E[Np(1-p)] + Var(Np)$$

which provides a *variance decomposition* of the marginal variance into a part reflecting the variance within a study $E[Np(1-p)]$ and a variance due to the heterogeneity of studies which is $Var(p) = \tau^2$. Note that in case of *homogeneity* where the distribution of p is reduced to a one-point distribution with all mass at p, the marginal variance reduces to the conventional binomial variance $Var(Y) = Np(1-p)$. We can develop (9.7) even more:

$$Var(Y) = N\mu - NE(p^2) + N^2\tau^2 = N[\mu(1-\mu) - \tau^2] + N^2\tau^2, \qquad (9.8)$$

where $E[p(1-p)] = \mu(1-\mu) - \tau^2$ is the population-averaged variance due to within study variation. Since $p(1-p) \geq 0$ for $0 \leq p \leq 1$, it follows that

$$E[p(1-p)] = \mu(1-\mu) - \tau^2 \geq 0$$

or, equivalently

$$\tau^2 \leq \mu(1-\mu) \leq \frac{1}{4},$$

where the inequality becomes sharp for $\mu = \frac{1}{2}$ (see also Figure 9.1). Let the distribution of p be concentrated with equal weights at 0 and 1 so that $Var(p) = \frac{1}{4}$ and all inequalities become sharp. This shows that the maximum value for τ^2 is achieved with $\tau^2 = \frac{1}{4}$ and suggests to define a *measure of heterogeneity* as follows:

$$\omega = 4 \times \tau^2. \qquad (9.9)$$

The measure ω as defined in (9.9) is easy to interpret since

- it attains the value zero if there is no heterogeneity ($\tau^2 = 0$) and
- it attains the value 1 if there is maximum heterogeneity ($\tau^2 = \frac{1}{4}$).

Alternatively, one could relate τ^2 to the total variance $Var(Y)$ and define

$$\Omega = \frac{N^2\tau^2}{Var(Y)} = \frac{N^2\tau^2}{[\mu(1-\mu) - \tau^2]N + N^2\tau^2}.$$

This measure has favorable properties as well including

- $0 \leq \Omega \leq 1$
- with the extremes being attained for $\tau^2 = 0$ and $\tau^2 = \mu(1-\mu)$.

In particular $\Omega \geq 0$, since $\mu(1-\mu) - \tau^2 \geq 0$.

One disadvantage of Ω is that it involves N which might be different from

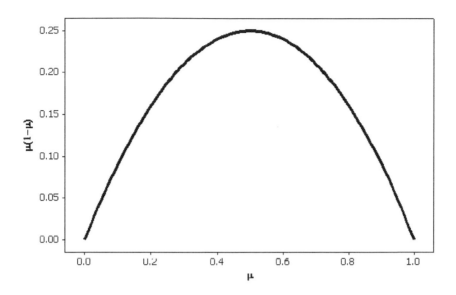

Figure 9.1 *Graph of $\mu(1 - \mu)$*

study to study and some way of allowing for a distribution of N must be incorporated. In case N is different from study to study we suggest replacing N by its expected values so that

$$\Omega = \frac{E(N^2)\tau^2}{[\mu(1 - \mu) - \tau^2]E(N) + E(N^2)\tau^2}$$

for the more general case.

9.3.1 Estimation

Estimates of μ and τ^2 must be provided. We suggest the Mantel-Haenszel estimator $\hat{\mu} = \frac{\sum_i y_i}{\sum_i N_i}$ which is known to behave well even in high sparsity cases. To estimation of τ^2 we consider (9.8) and conclude that

$$E(Y_i - N_i\mu)^2 = \qquad N_i\mu(1 - \mu) + N_i(N_i - 1)\tau^2, \text{ or} \qquad (9.10)$$

$$\sum_{i=1}^{k} E(Y_i - N_i\mu)^2 = \left(\sum_{i=1}^{k} N_i\right)\mu(1 - \mu) + \left(\sum_{i=1}^{k} N_i(N_i - 1)\right)\tau^2, (9.11)$$

so that a moment estimator for τ^2 can be constructed as

$$\hat{\tau}^2 = \frac{\sum_{i=1}^{k}(y_i - N_i\mu)^2 - \left(\sum_{i=1}^{k} N_i\right)\mu(1 - \mu)}{\sum_{i=1}^{k} N_i(N_i - 1)}. \qquad (9.12)$$

Clearly, the estimate $\hat{\tau}^2$ is *unbiased*, since $E(\hat{\tau}^2) = \tau^2$, but involves the unknown mean parameter μ which is replaced by $\mu = \frac{\sum_i y_i}{\sum_i N_i}$ so that an applicable version of (9.12) is provided as

$$\hat{\tau}^2 = \frac{\sum_{i=1}^{k}(y_i - N_i\hat{\mu})^2 - \left(\sum_{i=1}^{k} N_i\right)\hat{\mu}(1 - \hat{\mu})}{\left(\sum_{i=1}^{k} N_i(N_i - 1)\right)}.$$

Note that the estimator (9.12) is also defined in the extreme case $N_i = 0$ for some i which occurs when there are no events in both treatment arms. This is in contrast to the estimator defined in (Böhning (2000)) where (9.10) is rewritten as

$$\frac{E(Y_i - N_i\mu)^2}{N_i(N_i - 1)} = \frac{1}{N_i - 1}\mu(1 - \mu) + \tau^2,$$

so that the estimator follows,

$$\hat{\tau}_B^2 = \frac{1}{k}\sum_i \frac{(y_i - N_i\hat{\mu})^2}{N_i(N_i - 1)} - \frac{1}{k}\left(\sum_i \frac{1}{N_i - 1}\right)\hat{\mu}(1 - \hat{\mu}), \qquad (9.13)$$

which is also asymptotically unbiased but suffers from the obvious disadvantage of being *undefined* if $N_i \leq 1$. Similarly for the estimator

$$\hat{\tau}_M^2 = \frac{\sum_{i=1}^{k}(y_i - N_i\hat{\mu})^2/N_i - k\hat{\mu}(1 - \hat{\mu})}{\sum_{i=1}^{k}(N_i - 1)}, \qquad (9.14)$$

suggested and discussed in Lachin (2000)(p. 324-325) and Marshall (1991), where studies with $N_i = 0$ need to be removed (which seems to be no large restriction). Also, a number of estimators have been compared in Böhning et al. (2004), where it was found by empirical evidence that the estimator (9.14) behaved very well in many cases in comparison to others. Note that all estimators are truncated at 0 if they become negative.

9.3.2 Estimating the heterogeneity variance of θ

We have provided ways of estimating $Var(p) = \tau_p^2$, now indexed with p to avoid ambiguities in notation. However, original interest lies in estimating $Var(\theta) = \tau_\theta^2$. Now, *theta* and p are connected by means of $p = p(\theta) = \theta/(1+\theta)$ or inversely by $\theta = \theta(p) = p/(1 - p)$. We use the δ-method to calculate $Var(\theta)$. In summary, the δ-method provides a general technique to calculate the variance of a transformation $T(X)$ for a random variable X with given variance $Var(X)$. Use a first order Taylor series expansion around $E(X)$

$$T(X) \approx T(E(X)) + T'(E(X))(X - E(X))$$

and the result $Var(T(X)) \approx T'(E(X))^2 E(X - E(X))^2 = T'(E(X))^2 Var(X)$ follows. In our case, the transformation is $p/(1-p)$ with derivative $[p/(1-p)]' =$

$\frac{1}{(1-p)^2}$. Hence,

$$\tau_\theta^2 = Var(\theta) \approx \frac{1}{(1-\mu)^4} Var(p) = \frac{1}{(1-\mu)^4}\tau_p^2 \qquad (9.15)$$

and an estimator for τ_θ^2 is given as

$$\hat{\tau}_\theta^2 = \frac{\hat{\tau}_p^2}{(1-\hat{\mu})^4},$$

where we use for $\hat{\mu} = \frac{\sum_i y_i}{\sum_i N_i}$, and $\hat{\tau}_p^2$ is estimated by (9.12).

Frequently, interest is in $\phi = \log\theta$ so that an expression for τ_ϕ^2 is desired. Since $\log\theta = \log[p/(1-p)] = \log(p) - \log(1-p)$ we achieve

$$\tau_\phi^2 = Var(\phi) \approx \left(\frac{1}{\mu} + \frac{1}{(1-\mu)}\right)^2 Var(p) = \frac{1}{\mu^2(1-\mu)^2}\tau_p^2 \qquad (9.16)$$

and an estimator for τ_ϕ^2 is given as

$$\hat{\tau}_\phi^2 = \frac{\hat{\tau}_p^2}{\hat{\mu}^2(1-\hat{\mu})^2}.$$

An estimator for Ω can be found by plugging in estimates for μ and τ^2 and replacing $E(N)$ and $E(N^2)$ by their sample averages:

$$\hat{\Omega} = \frac{\left(\sum_i N_i^2\right)\hat{\tau}^2}{\left[\hat{\mu}(1-\hat{\mu}) - \hat{\tau}^2\right]\left(\sum_i N_i\right) + \left(\sum_i N_i^2\right)\hat{\tau}^2} \qquad (9.17)$$

9.3.3 Unbalanced trials

Although in many cases the trial arms are designed to be balanced, they turn out to be not balanced for a variety of reasons. In such situations the transformation $p = \frac{\theta n^T}{n^C + \theta n^T}$ will involve the different sizes of the trial arms and heterogeneity in p can no longer be entirely attributed to θ. To cope with this problem we suggest to *internally standardize* the trial to achieve equal trial arm sizes: $n_i = (n_i^T + n_i^C)/2$ with an associated "pseudo-number" of events $\tilde{x}_i^T = \frac{x_i^T}{n_i^T}n_i$ and $\tilde{x}_i^C = \frac{x_i^C}{n_i^C}n_i$. This procedure is similar to *indirect standardization* well-known in epidemiology and demography (see, for example Woodward (1999)) where expected counts are constructed on the basis of a common reference population. Note that we have chosen n_i such that their sum corresponds to the sum of the original sizes: $2n_i = n_i^T + n_i^C$, so that the available information appears to be neither diminished nor inflated, just differently allocated.

Table 9.2 *MAIPD for studying the effect of developing cold feet during treatment for hypertension (Hall et al. (1991)); data contain number of patients x_i^T, x_i^{C1} developing cold feet and number of patients recruited n_i^T, n_i^{C1} in the treatment (T) and first control arm (C1), respectively, as well as $n_i = (n_i^T + n_i^{C1})/2$ and pseudo-events $\tilde{x}_i^T$ and $\tilde{x}_i^{C1}$*

Center i	x_i^T	n_i^T	x_i^{C1}	n_i^{C1}	n_i	$\tilde{x}_i^T$	$\tilde{x}_i^{C1}$
1	3	13	5	14	13.5	3.11538	4.82143
2	2	3	0	4	3.5	2.33333	0.00000
3	0	3	0	3	3.0	0.00000	0.00000
4	1	4	1	4	4.0	1.00000	1.00000
5	1	4	3	5	4.5	1.12500	2.70000
6	0	2	1	1	1.5	0.00000	1.50000
7	0	6	1	6	6.0	0.00000	1.00000
8	1	2	0	1	1.5	0.75000	0.00000
9	0	4	1	4	4.0	0.00000	1.00000
10	0	3	1	3	3.0	0.00000	1.00000
11	1	1	0	1	1.0	1.00000	0.00000
12	0	8	2	8	8.0	0.00000	2.00000
13	1	4	0	4	4.0	1.00000	0.00000
14	0	2	0	2	2.0	0.00000	0.00000
15	0	3	0	2	2.5	0.00000	0.00000
16	0	3	1	4	3.5	0.00000	0.87500
17	0	1	0	2	1.5	0.00000	0.00000
18	0	12	0	12	12.0	0.00000	0.00000
19	1	2	0	1	1.5	0.75000	0.00000
20	0	9	5	6	7.5	0.00000	6.25000
21	0	2	0	1	1.5	0.00000	0.00000
22	0	2	0	1	1.5	0.00000	0.00000

9.3.4 Application to the sparse multicenter hypertension trial

Based on the last two columns of Table 9.2 we find that $\hat{\mu} = \frac{\sum_i y_i}{\sum N_i} = 0.3333$ with $N_i = \tilde{x}_i^T + \tilde{x}_i^{C1}$ and $y_i = \tilde{x}_i^T$. This corresponds to a relative risk of $\theta = \mu/(1-\mu) = 0.5$ which implies that the risk of developing cold feet during treatment is half the size as for the first control treatment. We use (9.12) to estimate τ^2 (with μ replaced by $\hat{\mu}$) and find a value of 0.0211, corresponding to $\hat{\omega} = 0.0843$, a minor value for the heterogeneity variance and associated heterogeneity measure.

We are now looking at the risk of developing cold feet for treatment versus the second control treatment. Based on the last two columns of Table 9.3 we find that $\hat{\mu} = \frac{\sum_i y_i}{\sum N_i} = 0.6809$ with $N_i = \tilde{x}_i^T + \tilde{x}_i^{C1}$ and $y_i = \tilde{x}_i^T$ which

Table 9.3 *MAIPD for studying the effect of developing cold feet during treatment for hypertension (Hall et al. (1991)); data contain number of patients x_i^T, x_i^{C2} developing cold feet and number of patients recruited n_i^T, n_i^{C2} in the treatment (T) and second control arm (C2), respectively, as well as $n_i = (n_i^T + n_i^{C2})/2$ and pseudo-events $\tilde{x}_i^T$ and $\tilde{x}_i^{C2}$*

Center i	x_i^T	n_i^T	x_i^{C2}	n_i^{C2}	n_i	$\tilde{x}_i^T$	$\tilde{x}_i^{C2}$
1	3	13	1	12	12.5	2.88462	1.04167
2	2	3	0	3	3.0	2.00000	0.00000
3	0	3	0	2	2.5	0.00000	0.00000
4	1	4	0	4	4.0	1.00000	0.00000
5	1	4	0	2	3.0	0.75000	0.00000
6	0	2	1	2	2.0	0.00000	1.00000
7	0	6	0	6	6.0	0.00000	0.00000
8	1	2	1	2	2.0	1.00000	1.00000
9	0	4	0	4	4.0	0.00000	0.00000
10	0	3	0	4	3.5	0.00000	0.00000
11	1	1	0	2	1.5	1.50000	0.00000
12	0	8	1	8	8.0	0.00000	1.00000
13	1	4	0	3	3.5	0.87500	0.00000
14	0	2	0	2	2.0	0.00000	0.00000
15	0	3	0	2	2.5	0.00000	0.00000
16	0	3	0	3	3.0	0.00000	0.00000
17	0	1	0	2	1.5	0.00000	0.00000
18	0	12	0	12	12.0	0.00000	0.00000
19	1	2	0	1	1.5	0.75000	0.00000
20	0	9	0	8	8.5	0.00000	0.00000
21	0	2	1	2	2.0	0.00000	1.00000
22	0	2	0	1	1.5	0.00000	0.00000

corresponds to a relative risk of $\hat{\theta} = \hat{\mu}/(1 - \hat{\mu}) = 2.1341$. Evidently, treatment establishes an elevated risk when compared to the second control treatment. We use (9.12) to estimate τ^2 (with μ replaced by $\hat{\mu}$) and find a value of 0 (a negative variance estimate is truncated to 0).

9.4 Testing for heterogeneity in a MAIPD

As has been discussed in Section 9.3.1 heterogeneity can be estimated involving $\sum_{i=1}^{k}(Y_i - N_i\hat{\mu})^2$ leading to various versions for estimating τ^2. How can the hypothesis of *homogeneity*, namely $\tau^2 = 0$, be tested? It is tempting to

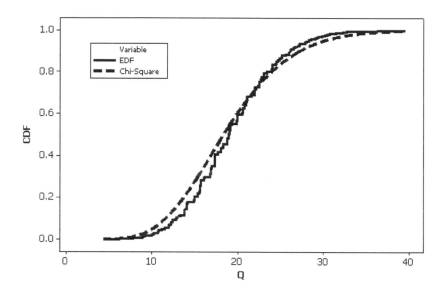

Figure 9.2 *Empirical distribution function and distribution of a χ^2 with 19 degrees of freedom*

consider

$$Q = \sum_{i=1}^{k} \frac{(Y_i - N_i\hat{\mu})^2}{N_i\hat{\mu}(1 - \hat{\mu})}, \tag{9.18}$$

where $\hat{\mu}$ is estimated by $\sum_i y_i / \sum_i N_i$. Q has been studied in detail by Potthoff and Whittinghill (1966). Usually, the approximation to a χ^2 with $(k-1)$ df is not good. However, they found that Q is well-behaved with increasing k in the sense that the χ^2−distribution can be used as an asymptotic distribution, even if the N_i remains small. However, for $k < 60$ the approximation is not good. We have simulated the distribution function of Q for $k = 20$ and N_i sampled from a distribution with mean $E(N_i) = 3$ with μ being chosen to be 0.3. The resulting empirical distribution function is shown in Figure 9.2 and is compared with the χ^2 with 19 degrees of freedom. Both distributions do not match exactly. The means agree reasonably well, but the true distribution of Q has a smaller variance than the comparable χ^2 with 19 df.

In coping with this problem Potthoff and Whittinghill (1966) refer to work by Nass (1959) who suggested a technique that seems to handle this latter problem and thereby provides the most accurate way of utilizing Q. Nass's refinement consists in taking

$$cQ \sim \chi^2_{\nu'} \tag{9.19}$$

where c and the fractional degrees of freedom ν' are chosen by Nass (1959)

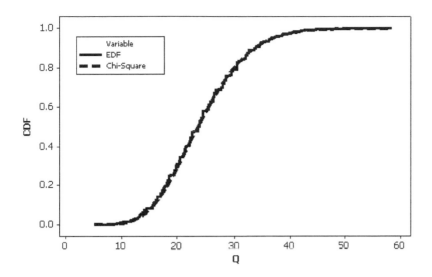

Figure 9.3 *Empirical distribution function and distribution of a χ^2 with 24.3 degrees of freedom*

such that $\chi^2_{\nu'}/c$ has the same first two moments as Q. Nass (1959) sets

$$c = 2E(Q)/Var(Q)$$

where $E(Q) = k - 1$ and $Var(Q) = 2(k-1) + \left(\frac{1}{\mu(1-\mu)} - 6\right) \sum_{i=1}^{k} \frac{1}{N_i}$. Then, under the null hypothesis, the statistic cQ follows approximately a $\chi^2_{\nu'}$ distribution with fractional degrees of freedom

$$\nu' = cE(Q) = 2[E(Q)]^2/Var(Q).$$

With these choices, we have that $Var(cQ) = 2E(cQ) = 2\nu'$. It remains to find estimates for $Var(Q)$ which can be simply given as

$$\widehat{Var(Q)} = 2(k-1) + \left(\frac{1}{\hat{\mu}(1-\hat{\mu})} - 6\right) \sum_{i=1}^{k} \frac{1}{N_i}.$$

With these modifications of Q we get excellent agreements of the distribution function of cQ and a χ^2-distribution with ν' degrees of freedom. Figure 9.3 shows both distribution functions for the simulation study as outlined above: $k = 20$, $\mu = 0.3$ and mean of the N_i being 3. The degrees of freedom were estimated from $\hat{\nu}' = \hat{c}E(Q) = 2[k-1]^2/\widehat{Var(Q)}$.

To illustrate the application of the test statistic let us consider again the sparse multicenter hypertension trial. We compare again treatment with the first control treatment. Note that there are seven centers with $N_i = 0$ which are not considered for this analysis, hence $k = 15$. We find that

$$Q = 18.88, \hat{c} = 2.59, \nu' = 36.23, \text{ and } \hat{c}Q = 48.85,$$

corresponding to a P-value of 0.08 which confirms our previous impression of minor heterogeneity in this MAIPD.

Finally, we compare treatment with the second control treatment. Note that there are 11 centers with $N_i = 0$ which are not considered for this analysis, hence $k = 11$. We find that

$$Q = 9.98, \hat{c} = 3.12, \nu' = 31.23, \text{ and } \hat{c}Q = 31.16,$$

corresponding to a P-value larger than 0.4699 which confirms that there is no heterogeneity in this MAIPD.

9.5 An analysis of the amount of heterogeneity in MAIPDs: a case study

In the following we will apply the developed concept of measuring heterogeneity to the MAIPDs previously used, in particular as presented in the eight Tables 1.5, 1.6, 1.7, 1.8, 1.9, 1.10, 1.11, and 1.12. The results including estimates of τ^2, Ω, and the value of $\hat{c}Q$ with associated P-value are provided in Table 9.4.

Table 9.4 *Assessment of the amount of heterogeneity in MAIPDs*

MAIPD	Amount of Heterogeneity			
Table	$\hat{\tau}_\theta^2$	$\hat{\Omega}$	$\hat{c}Q$	P-value
1.5	0.0022	0.1235	22.4940	0.4186
1.6	0.0046	0.5571	99.1111	0.0000
1.7	0.0000	0.0000	1.7487	0.9315
1.8	0.0000	0.0000	18.4277	0.8955
1.9	0.0152	0.4456	43.2867	0.0044
1.10	0.0137	0.4197	25.3067	0.0118
1.11	0.0000	0.0000	1.6955	0.7518
1.12	0.0000	0.0000	32.7539	0.5202

The MAIPDs in 1.7, 1.8, 1.11, and 1.12 show no heterogeneity at all which is supported by all Nass-modified χ^2-statistics which show all four MAIPDs having large P-values. For MAIPDs 1.6 and 1.10 we have τ^2-estimates of 0.0046 and 0.0137, respectively, which appear rather small. However, if the relative measure (Ω) is computed we find rather large estimates with values of 0.5571 and 0.4197 both accompanied by Nass-modified χ^2-statistics with small and significant P-values. This supports the importance of the relative heterogeneity measure Ω. Finally, we consider the MAIPDs 1.5 and 1.9 with estimates of τ^2 of 0.0022 and 0.0152, respectively. The first one has an Ω−value of 0.1235 with a nonsignificant and moderately large P-value, whereas the second one has an Ω−value of 0.4456, which is associated with a significant P-value indicating the presence of considerable heterogeneity.

9.6 A simulation study comparing the new estimate and the DerSimonian-Laird estimate of heterogeneity variance

In this section we compare the new estimate of heterogeneity variance with the DerSimonian-Laird estimate by means of a simulation study. The main settings of the simulation were described in detail in Chapter 5.5. In short, we generate the baseline risk p_i^C as uniformly distributed in the interval of 0.3 to 0.4. $p_i^T = \theta_j p_i^C$ depends on the chosen mixing distribution P and the index j denotes the component of the discrete distribution P which gives weight q_j to θ_j, whereas the index i denotes the study number. The sample sizes n_i^T and n_i^C are generated from a Poisson distribution with parameters μ_i^T and μ_i^C, respectively. The parameter $\mu_i^T = \mu_i^C$ was chosen from $\{5, 10, 20, 30, 40\}$. x_i^T is a binomial variate with parameters n_i^T and p_i^T, and x_i^C is a binomial variate with parameters n_i^C and p_i^C. The number of studies was set to be $k = 100$. The parameter $\log(\theta)$ can be considered as a random variable with distribution according to the mixing distribution P and its variance is readily available as

$$\tau^2 = Var(\log(\theta)) = \sum_{i=1}^{m} q_i(\log(\theta_i) - \overline{\log(\theta)})^2,$$

$$\text{with } \overline{\log(\theta)} = \sum_{i=1}^{m} q_i \log(\theta_i).$$

We use different values for heterogeneity variance, namely $\tau_1^2 = 0.3$, $\tau_2^2 = 0.5$, $\tau_3^2 = 0.8$, $\tau_4^2 = 1.0$. The corresponding mixing distributions are given in Table 9.5. For each condition we used 1,000 replications to compute the mean and variance of the two variance estimators.

The results of the simulation study in Figure 9.4 show that especially in the high sparsity case, if there are 5–10 participants per center in treatment or control arm, the new variance estimate has a *smaller* bias than the DerSimonian-Laird variance estimator. For large heterogeneity such as with a heterogeneity variance of 0.8–1.0, the DerSimonian-Laird estimator seems to be better than the new variance approach if there are more than 20 participants per center. The variance of both variance estimations is shown in Figure 9.5. Here it seems that the new variance estimator has higher variation if the number of participants per center in treatment or control arm is small. However, the low variance of the DerSimonian-Laird estimator is an *artificial effect*, because very frequently the DerSimonian-Laird estimator provides a negative variance

Table 9.5 *Heterogeneity and mixing distributions P with associated variances τ^2*

τ^2	P	
0.3	$\begin{pmatrix} 0.3 & 0.897154 \\ 0.5 & 0.5 \end{pmatrix}$	
0.5	$\begin{pmatrix} 0.3 & 1.23398 \\ 0.5 & 0.5 \end{pmatrix}$	
0.8	$\begin{pmatrix} 0.3 & 1.79478 \\ 0.5 & 0.5 \end{pmatrix}$	
1.0	$\begin{pmatrix} 0.3 & 2.21672 \\ 0.5 & 0.5 \end{pmatrix}$	

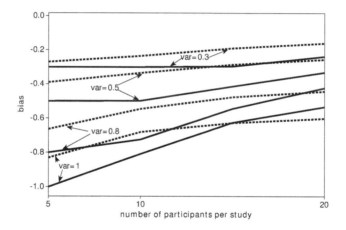

Figure 9.4 *Bias of the DerSimonian-Laird estimator (solid) and of the new variance estimator (dotted) computed by means of a simulation experiment for four values of heterogeneity variance var $= \tau^2$*

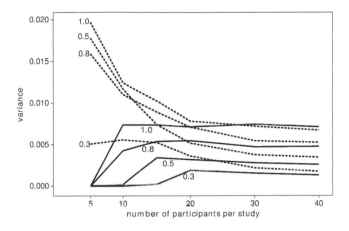

Figure 9.5 *Variance of the DerSimonian-Laird estimator (solid) and of the new vari-*
ance estimator (dotted) computed by means of a simulation experiment for four val-
ues of heterogeneity variance var = τ^2 as indicated with their values next to the lines
in the graph

value which needs to be truncated to zero. Indeed, with a mean of five participants per center the mean estimated variance of the DerSimonian-Laird
estimator goes to zero. This implies that in most of the simulation replications the estimated variance is set to zero (if the mean is exactly zero then all
variance estimates have to be zero) and the variation is consequently small.
This aspect is illustrated with the help of a simulation in which the nontrun-
cated variance estimators are computed (negative variances can occur here).
The mean of the variance estimator is presented for various combinations of
heterogeneity variance and number of participants per study. The results are
provided in Figure 9.6. The larger variance for the new variance estimator
and the more meaningful inverse relationship with increasing size per center
is an advantage of the new estimator. If the number of participants per center
increases, both variance estimators behave similarly.

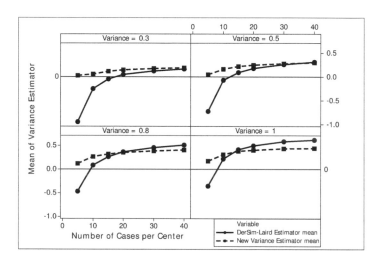

Figure 9.6 *Mean of the DerSimonian-Laird variance estimator (solid) and of the new variance estimator (dotted) computed by means of a simulation experiment for four values of heterogeneity variance var = τ^2 as indicated with their values next to the lines in the graph*

9.6.1 A simulation study comparing the conventional Q-test with the Nass-modified Q-test

Here we compare the modified Q-statistic (9.19) with the original and conventional Q-statistic (9.18). At first, we look at the null distribution. In Table 9.6 we report how often the two test statistics reject out of 1,000 replications.

From Table 9.6 we find the well-known result that the conventional Q-statistic does not behave like a χ^2 distributed random variable. Only when the individual study sizes increase considerably does the conventional Q-statistic start to reject occasionally. The Nass-modified Q-test does also not reach the desired rejection level of 50 in 1,000. However, it behaves superior to the conventional Q-test in all cases.

Now we look at the power of the two tests for the following scenario. The control arm has risk p_i^C as before. For the treatment arm, 50% of the observations arise from p_i^T, the other half from p_i^C. There are again k studies in the trial and $n_i^T = n_i^C$ participants in each trial arm. The results are shown in Table

Table 9.6 *Comparison of rejection proportions for both heterogeneity tests under homogeneity*

	Design Details			Rejection Counts out of 1,000	
k	$n_i^T (= n_i^C)$	p_i^T	p_i^C	cQ in (9.19)	Q in (9.18)
10	10	0.2	0.1	22	0
10	20	0.1	0.1	40	0
20	10	0.1	0.1	35	0
20	20	0.1	0.1	30	0
20	50	0.1	0.1	31	17
20	100	0.1	0.1	25	25
50	20	0.1	0.1	21	0

Table 9.7 *Comparison of rejection proportions for both heterogeneity tests under heterogeneity*

	Design Details			Rejection Counts out of 1,000	
k	$n_i^T (= n_i^C)$	p_i^T	p_i^C	cQ in (9.19)	Q in (9.18)
20	10	0.2	0.1	43	0
20	20	0.2	0.1	77	0
20	20	0.3	0.1	248	7
20	50	0.3	0.1	706	688
50	20	0.3	0.1	399	2
100	20	0.3	0.1	614	0

9.7. In all cases, the Nass-modified Q-test (9.19) has larger power than the conventional one (9.18). The latter appears to be without any power unless the number of participants per study becomes large. The remarkable benefit of (9.19) appears to be that it behaves well with increasing number of studies k while the number of participants per study, n_i^T and n_i^C, remains small. This makes it more suitable for sparse MAIPDs. Finally, these results are promising and illustrate a potential for a new and better test for homogeneity. However, we also feel that a deeper and more complete study on the behavior of the new test statistic is needed.

CHAPTER 10

Scrapie in Europe: a multicountry surveillance study as a MAIPD

10.1 The problem

In this chapter we would like to show that the methodology associated with the profile likelihood approach is not restricted to clinical trials but can be used successfully in other areas. Here, we consider a quite typical situation arising in disease surveillance with data arising from an observational study framework. The following introduction to the issue and problems of scrapie follows closely the recent work of Del Rio Vilas et al. (2007). Scrapie is a fatal neurological disease affecting small ruminants belonging to the group of diseases known as transmissible spongiform encephalopathies (TSE) that among others include bovine spongiform encephalopathy (BSE) in cattle and Creutzfeldt-Jakob disease (CJD) in humans. BSE was first detected in 1986 and was shown to spread between cattle by contaminated concentrate, see Wilesmith et al. (1988). In 1996 it became evident and likely that BSE transmits to humans and gives variant CJD (Will et al. (1996)). Throughout Europe, scrapie has acquired increased interest because it is considered a potential threat to public health after the successful experimental transmission of BSE to sheep, see Foster et al. (2001) and the likely exposure of sheep to concentrate feed contaminated with the BSE agent, see Hunter (2003). In order to obtain better estimates of the scrapie prevalence throughout the EU, active surveillance for scrapie in small ruminants was introduced in 2002. The surveillance comprised both slaughtered and found-dead animals, namely the abattoir (AS) and fallen stock (FS) surveys respectively, with the target numbers calculated for each country based on the adult sheep and goat populations (European Commission (2001)).

In 2003 the EU Commission Report on the monitoring for the presence of scrapie (European Commission (2004)) reported large variation in the frequency estimates of the two surveys between countries. In most of the countries the frequency estimates from the FS were larger than those of the AS. In other countries, however, the FS seemed to detect less scrapie than the AS. This pattern is inconsistent with other works that reported the increased risk of scrapie among the dead on farm animals (Wilesmith et al. (2004) and Del Rio Vilas et al. (2005)) and suggests the occurrence of heterogeneity in the

Table 10.1 *Comparison of the fallen stock and abattoir survey in their ability to detect scrapie*

Country	Fallen Stock		Abattoir S.		
	x_i^T	n_i^T	x_i^C	n_i^C	$\widehat{logRR}$
Belgium	2	494	0	2,376	3.18
Denmark	0	1,320	0	871	−0.42
Germany	13	48,616	9	20,107	−0.53
Greece	13	780	49	22,564	2.07
Spain	8	12,942	19	49,921	0.52
France	34	18,955	46	44,641	0.56
Ireland	18	2,830	9	51,579	3.57
Italy	13	5011	14	35260	1.88
Luxembourg	0	244	0	213	−0.14
Netherlands	6	3,994	45	21,095	−0.28
Austria	0	3,255	0	4,225	2.48
Portugal	0	243	6	10,697	1.22
Finland	0	683	0	1,990	1.07
Sweden	0	2,849	2	5,175	−1.01
UK	13	5,113	45	72,473	1.44
Czech Rep.	0	2,528	1	425	−2.88
Slovakia	1	213	1	3,923	2.91
Norway	8	3,359	5	33,519	2.74

implementation of the surveys between countries; surveys may be reflecting either different situations (e.g., different risks affecting the target individuals by the surveys, tests with different characteristics) or differences in their methodological implementation. There is a need to inform any comparisons between the detected prevalences in the individual surveillance streams. There have been previous attempts to inform these comparisons. Bird (2003) compared the surveillance performance of the two active surveillance sources among EU countries for BSE and scrapie in cattle and sheep respectively. Bird used the test results from 2001 and 2002, as reported by the EU Commission, to calculate the BSE and TSE rate ratios for each country to describe differences and anomalies in the implementation of the surveys. Bird also produced an EU-pooled measure of the rate ratios between surveillance streams: the median TSE prevalence rate ratio (fallen sheep vs. slaughtered). For the period April to August 2002, Bird reported a rate ratio of seven which indicated some conformity with the reported 10 times higher prevalence in the fallen stock group for cattle, see Scientific Steering Committee (2001). This increased "efficacy"of the FS is consistent with other works on sheep scrapie (Wilesmith et al. (2004) and Del Rio Vilas et al. (2005)). Following an approach similar to that of Bird (2003), the comparison of some form of risk ratio between the

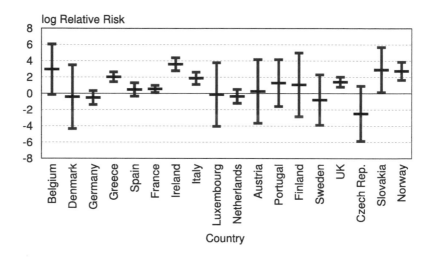

Figure 10.1 *Effect of type of survey (fallen stock vs. abattoir) for detecting scrapie by European country or 18 countries expressed as log-relative risk (AS is reference) with 95% confidence interval*

two surveys throughout the EU, under the standard conditions that apply to the surveys' operations, appears as an adequate methodology to assess the comparability of the scrapie surveillance across Europe. Exploring any differences in the ratios is important because it might help in the understanding of the performance of the surveillance programs.

The objective of this chapter is to demonstrate how the methodology developed so far for a MAIPD can be used to the apparent heterogeneity in the behavior of surveillance of scrapie across Europe and to investigate the sources of this heterogeneity by taking into account available country-level covariates.

10.2 The data on scrapie surveillance without covariates

Data on the number of sheep tested and confirmed by each surveillance source (AS and FS) were collected from the EU's annual report on the monitoring of transmissible spongiform encephalopathy (TSE) in ruminants in 2003, see European Commission (2004). Data from 18 European countries were available for the study (see Table 10.1). The number x^T (x^C) of positive samples out of n^T (n^C) samples tested for the fallen stock survey (abattoir survey) are listed

Table 10.2 *Mixture model assessment for the data on scrapie in Europe: no covariates considered*

Number of components	Log-Likelihood	BIC	Gradient function
1	−4069.1	−8141.1	
2	−4041.0	−8090.7	2.7476
3	−4034.9	−8084.2	1.3253
4	−4033.5	−8087.2	1

in Table 10.1 for each European country. If p^T denotes the risk of scrapie detection in the FS and if p^C denotes the risk of scrapie detection in the AS, then the rate ratio $RR = p^T/p^C$ expresses the amount of ability with which the FS can better detect scrapie than the AS. For example, if $RR = 2$ there is two times higher ability to detect scrapie for the FS in comparison to the AS. Hence, the rate ratio has become a common effect measure in this field. The RR is simply estimated as $\widehat{RR} = \frac{x^T/n^T}{x^C/n^C}$. It is not surprising that there is considerable variation of the $\widehat{RR}$ across Europe (see also Figure 10.1).

10.3 Analysis and results

As can be seen from the last column of Table 10.1 the logarithmic risk ratios, relating the detection risk of the FS surveillance source to the detection risk of the AS surveillance source, are experiencing considerable heterogeneity. In the following we will try to explore this heterogeneity in more detail.

To avoid computational abnormalities, we have added 0.5 only if $x^T = 0$ to the fallen stock's arm or if $x^C = 0$ to the abattoir stock's arm. For the data in Table 10.1 we find a measure of heterogeneity Ω (developed in Section 9.3) as 0.939, indicating a clear presence of heterogeneity in the effect. The overall relative risk is 3 with 95% confidence interval of (2.358–3.793) under the assumption of homogeneity. This result indicates a clear advantage for the fallen stock survey.

However, as the high measure of heterogeneity indicates, we need to consider modeling heterogeneity. Using the profile mixture likelihood approach, the EMGFU algorithm finds a global maximum of four mixture components. The result is shown in Table 10.3 and the classification into the mixture components is presented in Table 10.4. Using the Bayesian Information Criterion as a model selection device three-mixture components appear more appropriate, as Table 10.2 shows. The details for the three-component mixture model are provided in Table 10.5.

It is worth noting that in the three-component solution the estimate for the second mixtures component is nonsignificant. In this component are most of

Table 10.3 *Results for the scrapie MAIPD with four mixture components*

Comp.	Weights	N	Cov.	$\hat{\beta}_i$	S.E.	P-value
1	0.1221	1	Intercept	3.423	0.412	0.000
2	0.2419	2	Intercept	0.513	0.234	0.014
3	0.3775	9	Intercept	1.804	0.202	0.000
4	0.2585	6	Intercept	−0.393	0.350	0.131
Log-Likelihood	−4033.5					
BIC	−8087.2					
Max of G.-function	1.0000					

the countries. Ireland is classified as the only country in component one with a relative risk of 30. The third component classified six countries with a relative risk of six.

Even the four-component solution has one component with six countries where the relative risk is not significant. France, Spain, Austria, Finland, and Portugal change from the nonsignificant relative risk component to a significant component.

In summary and for the overall picture, it seems that there is sufficient evidence that the data can be classified into three components (see Table 10.6):

- a majority cluster consisting of 11 countries with a slightly increased, but nonsignificant risk ratio. Here the FS seems to be able not to detect a lot more scrapie than the AS.
- a cluster consisting of six countries with significantly increased risk ratio. In this cluster the ability for detecting scrapie is six times as high for the FS in comparison to the AS.
- and, finally, there is one country cluster consisting of Ireland with a largely increased risk ratio.

10.4 The data with covariate information on representativeness

Additional covariate information was available in form of the proportion of the country population sampled by the fallen stock (*RP-FS*) and the proportion of the population sampled by the abattoir survey (*RP-AS*) which were taken as measures of representativeness in Del Rio Vilas et al. (2007). The additional data are provided in Table 10.7.

As Figure 10.2 indicates, it cannot be excluded that these two covariates are

Table 10.4 *Classification of the countries into the four mixture components*

Mixture Component	Trial	$\widehat{RR}$ (95% CI)
1	Ireland	30.6620 (13.6733–68.7588)
2	France Spain	1.6702 (1.0559–2.6419)
3	Austria Belgium Finland Greece Italy Norway Portugal Slovakia UK	6.0738 (4.0858–9.0291)
4	Czech Rep. Denmark Germany Luxembourg Netherlands Sweden	0.6748 (0.3397–1.3405)

Table 10.5 *Results for the scrapie MAIPD with three mixture components*

Comp.	Weights	N	Cov.	$\hat{\beta}_i$	S.E.	P-value
1	0.1216	1	Intercept	3.423	0.412	0.000
2	0.4997	11	Intercept	0.219	0.169	0.097
3	0.3787	6	Intercept	1.798	0.198	0.000
Log-Likelihood			−4034.9			
BIC			−8084.2			
Max of G.-function			1.3253			

associated with the log-relative risk. Hence, they must be considered appropriately as potential covariates in the modeling.

Table 10.6 *Classification of the countries into the three mixture components*

Component	Trial	$\widehat{RR}$ (95% CI)
1	Ireland	30.6556 (13.6839–68.6770)
2	Austria Czech Rep. Denmark Finland France Germany Luxembourg Netherlands Portugal Spain Sweden	1.2449 (0.8942–1.7330)
3	Belgium Greece Italy Norway Slovakia UK	6.0363 (4.0967–8.8944)

10.4.1 Analysis of data on scrapie in Europe under incorporation of covariates

We consider first the more general model that allows mixing simultaneously on intercept and the two slopes for the two covariates. In Chapter 6 the following mixture model was discussed:

$$\log \theta_j = \eta_j = \beta_{0j} + \beta_{1j} z_1 + \beta_{2j} z_2.$$

This is equation (6.2) with the two ($p = 2$) covariates for representativeness. The index j is the mixture component and mixing goes from $j = 1, ..., m$ where m is the number of mixture components. Note that this model has coefficients for the covariates that are allowed to vary from mixture component to mixture component. Table 10.8 provides an evaluation of the mixture model from one to four components, the latter being the NPMLE. There is a clear support for the three-component model on the basis of the BIC. Table 10.9 provides details on the model estimates for the three-component mixture model. Note that there is considerable variation in the slope estimates for the three components.

Table 10.10 shows the classification of the countries in Europe into the three-mixture components using the MAP allocation rule. Finally, it might be valu-

Table 10.7 *Comparison of the fallen stock and abattoir survey in their ability to detect scrapie*

Country	Fallen Stock		Abattoir S.		Repres.	
	x_i^T	n_i^T	x_i^C	n_i^C	RP-FS	RP-AS
Belgium	2	494	0	2,376	0.34	1.63
Denmark	0	1,320	0	871	1.26	0.83
Germany	13	48,616	9	20,107	1.84	0.76
Greece	13	780	49	22,564	0.01	0.25
Spain	8	12,942	19	49,921	0.06	0.22
France	34	18,955	46	44,641	0.21	0.50
Ireland	18	2,830	9	51,579	0.05	0.87
Italy	13	5011	14	35260	0.06	0.44
Luxembourg	0	244	0	213	3.49	3.04
Netherlands	6	3,994	45	21,095	0.31	1.66
Austria	0	3,255	0	4,225	1.07	1.39
Portugal	0	243	6	10,697	0.01	0.31
Finland	0	683	0	1,990	1.02	2.97
Sweden	0	2,849	2	5,175	0.63	1.15
UK	13	5,113	45	72,473	0.02	0.30
Czech Rep.	0	2,528	1	425	2.45	0.41
Slovakia	1	213	1	3,923	0.07	1.21
Norway	8	3,359	5	33,519	0.36	3.61

Table 10.8 *Mixture model assessment for the data on scrapie in Europe under inclusion of two covariates (RP-FS and RP-AS)*

Number of components	Log-Likelihood	BIC	Gradient function
1	−4054.4	−8117.5	
2	−4030.9	−8076.2	5.7074
3	−4026.2	−8072.6	1.1569
4	−4026.1	−8078.6	1

able to consider the four-component mixture model as an alternative since it is based upon the nonparametric maximum likelihood. Here, the maximum of gradient function attains the value one, so that not more than four components can exist. Table 10.11 contains the details of the model fit for the four-component mixture model and Table 10.12 shows the allocation of the European countries into the four-mixture components using the MAP allocation rule. Figure 10.3 contains a scatterplot of the observed log-relative risk against the fitted values of the log-relative risk for the three and four components. Both fits appear reasonable with the only difference between the two

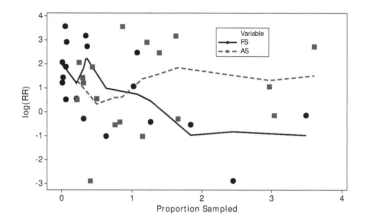

Figure 10.2 *Scatterplot of log-relative risk and proportion sampled for fallen stock survey and abattoir survey (with LOWESS-smoother for each group)*

models being in the two countries Belgium and Slovakia. This might provide further evidence why the BIC criterion selected the three-component model.

Table 10.9 *Results of the MAIPD on scrapie with three-mixture components (BIC criterion) and included covariates RP-FS and RP-AS*

Comp.	Weights	S.E. of W	N	Cov.	$\hat{\beta}_i$	S.E.	P-value
1	0.2673	.1725	3	Intercept	2.3032	0.4157	0.0000
				RP-FS	−1.0165	0.5113	0.0234
				RP-AS	−1.3682	0.4164	0.0005
2	0.2185	.1725	2	Intercept	0.4450	0.8591	0.3022
				RP-FS	−2.0532	0.5313	0.0001
				RP-AS	3.5852	1.3400	0.0037
3	0.5142		13	Intercept	0.4276	0.2833	0.0656
				RP-FS	−0.9131	0.3400	0.0036
				RP-AS	0.7549	0.1897	0.0000
Log-Likelihood	−4026.2						
BIC	−8072.6						
Max of G.-function	1.1569						

10.4.2 Analysis of scrapie in Europe with covariates in a simplified model

We consider the general model that allows mixing simultaneously on intercept and the two slopes for the two covariates. In Chapter 6 the following *simplified* mixture model was discussed:

$$\log \theta_j = \eta_j = \beta_{0j} + \beta_1 z_1 + \beta_2 z_2,$$

where z_1 and z_2 are the two covariates for representativeness. The index j for the mixture component is now dropped for the slope of the covariates of interest and mixing takes place only over the intercept β_{0j}. Note that this model assumes a *common* coefficient for each covariate in the model. Table 10.13 provides an evaluation of the mixture model from one to four components, the latter being again the NPMLE. There is a clear support for the three-component model on the basis of the BIC.

Table 10.14 provides details on the model estimates for the three-component mixture model. Table 10.15 gives an allocation of the countries in Europe into the three components.

Again finally, it might be valuable to consider the four-component mixture model as an alternative since it is based upon the NPMLE. It is the most complex model achievable in this simplified class. Table 10.16 contains the details of the model fit for the four-component mixture models, and Table 10.17 shows the allocation of the European countries into the four-mixture components using the MAP allocation rule.

Table 10.10 *Classification of the countries in Europe into the three-mixture components (BIC criterion) using a mixture model that incorporates the covariates RP-FS and RP-AS*

Mixture Component	Trial	$\widehat{RR}$ (95% CI)
3	Belgium	3.8480 (2.3938–6.1855)
3	Denmark	0.9081 (0.4650–1.7736)
3	Germany	0.5072 (0.1835–1.4023)
1	Greece	7.0353 (3.5684–13.8703)
3	Spain	1.7141 (1.0406–2.8233)
3	France	1.8464 (1.2084–2.8214)
2	Ireland	31.8624 (13.2397–76.6794)
1	Italy	5.1559 (2.8641–9.2814)
3	Luxembourg	0.6286 (0.0768–5.1458)
1	Netherlands	0.7533 (0.3230–1.7571)
3	Austria	1.6485 (0.9282–2.9277)
3	Portugal	1.9202 (1.1630–3.1706)
3	Finland	5.6870 (2.3323–13.8668)
3	Sweden	2.0553 (1.3716–3.0799)
2	UK	4.3906 (1.6587–11.6221)
3	Czech Rep.	0.2231 (0.0533–0.9341)
3	Slovakia	3.5860 (2.2350–5.7538)
3	Norway	16.8428 (5.6939–49.8214)

Figure 10.4 contains a scatterplot of the observed log-relative risk against the fitted values of the log-relative risk for the three-component models in full and simplified version. Both fits appear reasonable. The simplified model with common slope for all mixture components might be preferable here since it has fewer parameters and is easier to interpret and communicate.

In summary, we find that

- the proportion of sampled sheep in the fallen stock survey is *negatively* related to the ability ratio in detecting scrapie.

- the proportion of sampled sheep in the abattoir survey is *positively* related to the ability ratio in detecting scrapie.

This tendency has been visible already in Figure 10.1 and is now confirmed. In addition, we have seen that European countries can be grouped into three levels according to their detection ability ratio: a few countries with high relative risk, many countries with a moderately enlarged relative risk, and one with a decreased relative risk value.

Table 10.11 *Results of the MAIPD on scrapie using a mixture model with four components (NPMLE criterion) and included covariates RP-FS and RP-AS*

Comp.	Weights	S.E. of W	N	Cov.	$\hat{\beta}_i$	S.E.	P-value
1	0.1900	0.1666	1	Intercept	2.2456	0.7068	0.0007
				RP-FS	−0.9925	0.6362	0.0594
				RP-AS	−1.3559	0.5208	0.0046
2	0.1658	0.1512	2	Intercept	0.2730	0.8821	0.3785
				RP-FS	−2.0588	0.6026	0.0003
				RP-AS	3.8598	1.3008	0.0015
3	0.3761	0.2324	11	Intercept	0.3965	0.2820	0.0798
				RP-FS	−0.8992	0.3990	0.0121
				RP-AS	0.7587	0.2410	0.0008
4	0.2681		4	Intercept	1.6861	0.5355	0.0008
				RP-FS	−1.4064	0.4914	0.0021
				RP-AS	0.4626	0.3055	0.0650

Log-Likelihood	−4026.3
BIC	−8078.6
Max of G.-function	1.0000

Table 10.12 *Classification of the countries in Europe into the four-mixture components (NPMLE criterion) with incorporated covariates RP-FS and RP-AS*

Mixture Component	Trial	$\widehat{RR}$(95% CI)
4	Belgium	7.1126 (3.1664–15.9768)
3	Denmark	0.8988 (0.4068–1.9862)
3	Germany	0.5060 (0.1502–1.7050)
4	Greece	5.9753 (2.2720–15.7152)
3	Spain	1.6645 (1.0219–2.7110)
3	France	1.7987 (1.1923–2.7135)
2	Ireland	34.0577 (14.7559–78.6080)
4	Italy	6.0812 (2.4811–14.9054)
3	Luxembourg	0.6473 (.0540–7.7543)
1	Netherlands	0.7313 (.3125–1.7113)
3	Austria	1.6308 (0.8173–3.2542)
3	Portugal	1.8641 (1.1417–3.0435)
3	Finland	5.6568 (1.8104–17.6754)
3	Sweden	2.0190 (1.2714–3.2062)
2	UK	4.0138 (1.4123–11.4073)
3	Czech Rep.	0.2242 (0.0403–1.2457)
4	Slovakia	8.5619 (3.7262–19.6733)
3	Norway	16.6416 (4.0435–68.4911)

Table 10.13 *Mixture model assessment for the data on scrapie in Europe under inclusion of two covariates (RP-FS and RP-AS) in the simplified model*

Number of components	Log-Likelihood	BIC	Gradient function
1	−4054.4	−8117.5	
2	−4034.5	−8083.5	2.3605
3	−4031.1	−8082.5	1.1973
4	−4029.2	−8084.3	1

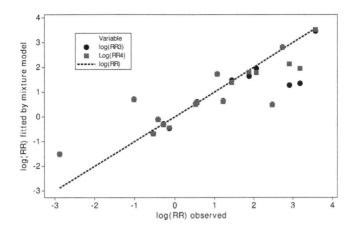

Figure 10.3 *Scatterplot of the fitted log-relative risk (log(RR3) is for the three-component model, log(RR4) is for the four-component model) against the observed log-relative risk*

Table 10.14 *Details of model estimation for the MAIPD on scrapie for a mixture model with three-components (BIC criterion) using the simplified model and including the covariates RP-FS and RP-AS*

Comp.	Weights	S.E. of W	N	Cov.	$\hat{\beta}_i$	S.E.	P-value
1	0.4484	.1736	7	Intercept	2.0350	0.2635	0.0000
				RP-FS	−1.7369	0.2647	0.0000
				RP-AS	0.8415	0.1993	0.0000
2	0.4667	.1770	10	Intercept	0.6558	0.2325	0.0024
				RP-FS	−1.7369	0.2647	0.0000
				RP-AS	0.8415	0.1993	0.0000
3	0.0849		1	Intercept	−1.1694	0.5374	0.0148
				RP-FS	−1.7369	0.2647	0.0000
				RP-AS	.8415	0.1993	0.0000

Log-Likelihood	−4031.1
BIC	−8082.5
Max of G.-function	1.1973

Table 10.15 *Classification of the European countries into the three-mixture compo-*
nents (BIC criterion) using the simplified model and including the covariates RP-FS
and RP-AS

Mixture Component	Trial	$\widehat{RR}$(95% CI)
1	Belgium	16.7128 (11.9195–23.4336)
2	Denmark	0.4342 (0.2474–0.7620)
1	Germany	0.5938 (0.2579–1.3669)
1	Greece	9.2815 (5.9481–14.4829)
2	Spain	2.0891 (1.4401–3.0304)
2	France	2.0377 (1.5595–2.6626)
1	Ireland	14.5894 (10.5380–20.198 3)
1	Italy	9.9848 (6.7559–14.7569)
1	Luxembourg	0.2303 (.0447–1.1874)
3	Netherlands	0.7328 (0.3123–1.7199)
2	Austria	0.9676 (0.6419–1.4587)
2	Portugal	2.4579 (1.7446–3.4628)
2	Finland	3.9891 (1.9280–8.2536)
2	Sweden	1.6978 (1.3904–2.0732)
2	UK	2.3953 (1.6957–3.3836)
2	Czech Rep.	0.0386 (0.0113–.1324)
1	Slovakia	18.7591 (13.5885–25.8972)
2	Norway	21.5088 (8.3750 55.2307)

Table 10.16 *Details of model estimation for the MAIPD on scrapie for a mixture model with four-components (NPMLE criterion) using the simplified model and including the covariates RP-FS and RP-AS*

Comp.	Weights	S.E. of W	N	Cov.	$\hat{\beta}_i$	S.E.	P-value
1	0.0888	0.0855	1	Intercept	3.1586	0.4707	0.0000
				RP-FS	−1.3515	0.2721	0.0000
				RP-AS	0.5354	0.2692	0.0234
2	0.2601	0.1814	3	Intercept	0.5997	0.2362	0.0056
				RP-FS	−1.3515	0.2721	0.0000
				RP-AS	0.5354	0.2692	0.0234
3	0.0918	0.0880	1	Intercept	−0.7726	0.6338	0.1114
				RP-FS	−1.3515	0.2721	0.0000
				RP-AS	0.5354	0.2692	0.0234
4	0.5593		13	Intercept	1.5734	0.2505	0.0000
				RP-FS	−1.3515	0.2721	0.0000
				RP-AS	0.5354	0.2692	0.0234
Log-Likelihood	−4029.2						
BIC	−8084.3						
Max of G.-function	1.0000						

Table 10.17 *Classification of the countries into the four-mixture components (NPMLE criterion) using the simplified model and including the covariates RP-FS and RP-AS*

Mixture Component	Trial	$\widehat{RR}$ (95% CI)
4	Belgium	7.2903 (3.8268–13.8885)
4	Denmark	1.3701 (0.7676–2.4456)
4	Germany	0.6026 (0.2589–1.4026)
4	Greece	5.4398 (3.5461–8.3448)
2	Spain	1.8896 (1.2485–2.8599)
2	France	1.7924 (1.2162–2.6416)
1	Ireland	35.0493 (15.0026–81.8824)
4	Italy	5.6287 (3.7979–8.3422)
4	Luxembourg	0.2196 (0.0335–1.4385)
3	Netherlands	0.7387 (0.2425–2.2505)
4	Austria	2.3904 (1.3138–4.3494)
4	Portugal	5.6174 (3.6985–8.5318)
4	Finland	5.9590 (1.6915–20.9933)
2	Sweden	1.4389 (0.8450–2.4502)
4	UK	5.5124 (3.6309–8.3687)
4	Czech Rep.	0.2191 (0.0666–0.7207)
4	Slovakia	8.3863 (4.9741–14.1393)
4	Norway	20.4817 (4.0534–103.4946)

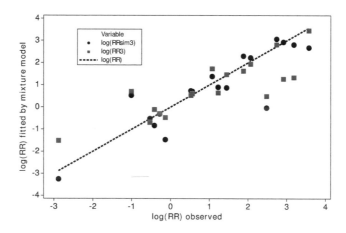

Figure 10.4 *Scatterplot of the fitted log-relative risk for the full three-component model (log RR3) and for the simplified three-component model against the observed log-relative risk*

A.1 Derivatives of the binomial profile likelihood

We provide details involved in the derivation of the function:

$$
L_{BIN}(\theta) = \sum_{i=1}^{k} x_i^T \log(\theta_i) + (n_i^T - x_i^T) \log(1 - \theta_i p_i^C(\theta_i))
$$
$$
+ \left(x_i^T + x_i^C \right) \log \left(p_i^C(\theta_i) \right) + \left(n_i^C - x_i^C \right) \log(1 - p_i^C(\theta_i)), \quad \text{(A.1)}
$$

where the derivation is given as

$$
\frac{\partial}{\partial \theta} L_{BIN}(\theta) = \sum_{i=1}^{k} \frac{x_i^T}{\theta} + \frac{\left(x_i^T + x_i^C \right) \left(p_i^C \right)'(\theta)}{p_i^C(\theta)} - \frac{\left(n_i^C - x_i^C \right) \left(p_i^C \right)'(\theta)}{1 - p_i^C(\theta)}
$$
$$
- \frac{(n_i^T - x_i^T)(p_i^C(\theta) + \theta \left(p_i^C \right)'(\theta))}{1 - \theta p_i^C(\theta)} \quad \text{(A.2)}
$$

The derivative of $p_i^C(\theta)$ is essential to compute the fixed-point iteration (5.26). The first derivative of $p_i^C(\theta)$ takes the following form:

$$
\frac{\partial}{\partial \theta} p_i^C(\theta) = \frac{\left(n_i^C \right)^2 + \left(x_i^T \right)^2 - \theta \left(2n_i^T x_i^C + n_i^T x_i^T - x_i^T x_i^C \right)}{2\theta^2 n_i w_i(\theta)}
$$
$$
= \frac{n_i^C \left(\theta n_i^T + 2x_i^T - \theta(x_i^C + 2x_i^T) \right)}{2\theta^2 n_i w_i(\theta)} - \frac{\left(n_i^C + x_i^T \right)}{2\theta^2 n_i}.
$$

The variance estimation of θ is based upon the second derivative of (A.1), given as

$$
\frac{\partial^2}{\partial \theta^2} L_{BIN}(\theta) = \sum_{i=1}^{k} \frac{x_i^T}{\theta^2} + \left(x_i^T + x_i^C \right) \left(-\frac{\left(p_i^C \right)'(\theta)^2}{p_i^C(\theta)^2} + \frac{\left(p_i^C \right)''(\theta)}{p_i^C(\theta)} \right)
$$
$$
- \left(n_i^C - x_i^C \right) \left(\frac{\left(p_i^C \right)'(\theta)^2}{\left(1 - p_i^C(\theta) \right)^2} + \frac{\left(p_i^C \right)''(\theta)}{1 - p_i^C(\theta)} \right)
$$
$$
- \left(n_i^T - x_i^T \right) \left(\frac{(-p_i^C(\theta) - \theta \left(p_i^C \right)'(\theta))^2}{\left(1 - \theta p_i^C(\theta) \right)^2} - \frac{-2 \left(p_i^C \right)'(\theta) - \theta \left(p_i^C \right)''(\theta)}{1 - \theta p_i^C(\theta)} \right).
$$
$$
\text{(A.3)}
$$

The second derivative of $p_i^C(\theta)$ is of the following form:

$$\frac{\partial^2}{\partial\theta^2}p_i^C(\theta) = \frac{1}{\left(n_i^C + n_i^T\right)\theta^3}\left(n_i^C + \theta\left(n_i^T + x_i^C\right) + x_i^T - w_i(\theta)\right.$$

$$- \frac{2n_i\theta^2\left(n_i^C - x_i^C\right)\left(n_i^T - x_i^T\right)x_i}{w_i(\theta)^3}$$

$$- \theta(n_i^T + x_i^C + \frac{-n_i^C n_i^T - \left(n_i^T\right)^2\theta + n_i^C x_i^C + 2n_i^T x_i^C}{w_i(\theta)}$$

$$+ \frac{-2n_i^T\theta x_i^C - \theta\left(x_i^C\right)^2 + 2n_i^C x_i^T + n_i^T x_i^T - x_i^C x_i^T}{w_i(\theta)})$$

$$\text{with } w_i(\theta) = \sqrt{\left(n_i^C + \theta\left(n_i^T + x_i^C\right) + x_i^T\right)^2 - 4\theta x_i n_i}.$$

A.2 The lower bound procedure for an objective function with a bounded Hesse matrix

In section 6.1 a modification of the conventional Newton-Raphson procedure was mentioned that we would like to study now in more detail. Consider an objective function $L(\beta)$, which is supposed to be maximized in β. Let $\nabla L(\beta)$ denote the *gradient*, the vector of first derivatives. If the second derivate matrix $\nabla^2 L(\beta)$ of $L(\beta)$ is bounded below by some matrix $\mathbf{B}$

$$\nabla^2 L(\beta) \geq \mathbf{B}, \tag{A.4}$$

where $\mathbf{A} \geq \mathbf{C}$ means that $\mathbf{A} - \mathbf{C}$ is nonnegative definite, then the iteration

$$\beta^{new} - \beta^0 - \nabla^2 L(\beta^0)^{-1}\nabla L(\beta^0) \tag{A.5}$$

will have the property that

$$L(\beta^{new}) \geq L(\beta^0). \tag{A.6}$$

This is because we have in the second order expansion for any β and some $\beta^* = (1 - \alpha)\beta_0 + \alpha\beta$ and $\alpha \in [0, 1]$

$$L(\beta) = L(\beta^0) + \nabla L(\beta^0)'(\beta - \beta^0) + (\beta - \beta^0)'\nabla^2 L(\beta^*)(\beta - \beta^0) \tag{A.7}$$

$$\geq L(\beta^0) + \nabla L(\beta^0)'(\beta - \beta^0) + (\beta - \beta^0)'\mathbf{B}(\beta - \beta^0).$$

The right-hand side of the inequality (A.7) is maximized for β given in (A.5). Hence we have the monotonicity property (A.6). More details are provided in Böhning and Lindsay (1988), Böhning (1992), Böhning (2000), or Lange (2004).

A.2.1 The lower bound procedure for the profile log-likelihood

In chapter 4, the following second derivative matrix of the profile log-likelihood was found in section 4.2:

$$\frac{\partial^2 L^*}{\partial \beta_h \partial \beta_j}(\beta) = -\sum_{i=1}^{k} \frac{x_i n_i^T n_i^C \exp(\eta_i)}{(n_i^C + \exp(\eta_i) n_i^T)^2} z_{ij} z_{ih} \qquad (A.8)$$

so that (A.8) becomes in matrix form:

$$\nabla^2 L^*(\beta) = \left(\frac{\partial^2 L^*}{\partial \beta_h \partial \beta_j}(\beta) \right) = -Z'W(\beta)Z \qquad (A.9)$$

where **Z** is the design matrix of covariate information defined as:

$$Z = \begin{pmatrix} z_{10} & z_{11} & z_{12} & \cdots & z_{1p} \\ z_{20} & z_{21} & z_{22} & \cdots & z_{2p} \\ \cdot & \cdot & \cdot & \cdots & \cdot \\ z_{k0} & z_{k1} & z_{k2} & \cdots & z_{kp} \end{pmatrix}_{k \times (p+1)}$$

with

k is the number of centers,

p is the number of covariates,

z_{i0} is the constant value of coefficient in the i-th center, $z_{i0} = 1$,

$z_{i1}, \ldots, z_{ip}$ is value of covariates in the i-th center,

and $W(\beta)$ is a diagonal matrix defined as:

$$W(\beta) = \begin{pmatrix} w_{11} & w_{12} & w_{13} & \cdots & w_{1k} \\ w_{21} & w_{22} & w_{23} & \cdots & w_{2k} \\ \cdot & \cdot & \cdot & \cdots & \cdot \\ w_{k1} & w_{k2} & w_{k3} & \cdots & w_{kk} \end{pmatrix}_{k \times k}$$

with $w_{ij} = 0$, if $i \neq j$ and

$$w_{ii} = \frac{x_i n_i^T n_i^C \exp(\eta_i)}{(n_i^C + \exp(\eta_i) n_i^T)^2}. \qquad (A.10)$$

To show that (A.9) has a global lower bound it is sufficient to prove that (A.10) is bounded above. We show (and ignore indeces for simplification)

$$\frac{n^T n^C \exp(\eta)}{(n^C + \exp(\eta) n^T)^2} \leq \frac{1}{4}. \qquad (A.11)$$

We do this by writing the left-hand side of (A.11) as

$$\frac{n^T n^C \exp(\eta)}{(n^C + \exp(\eta) n^T)^2} = p(1-p)$$

with $p = \frac{n^T \exp(\eta)}{n^C + \exp(\eta) n^T}$ and note that the proof is complete since $p(1-p)$ is the variance of a Bernoulli variable that is maximized for $p = 0.5$.

A.3 Connection between the profile likelihood odds ratio estimation and the Mantel-Haenszel estimator

The iteration (8.11) in Section 8.1:

$$
\Gamma_{OR}(\kappa) \quad := \quad \frac{\sum_{i=1}^{k} \frac{(n_i^C - x_i^C)x_i^T}{\sqrt{r_i(\kappa)}}}{\sum_{i=1}^{k} t_i(\kappa)} \tag{A.12}
$$

$$
\text{where } t_i(\kappa) \quad = \quad \frac{n_i^T \left(\kappa \times \left(q_i^C\right)'[\kappa] + q_i^C[\kappa]\right)}{1 + \kappa \times q_i^C[\kappa]} + \frac{n_i^C \left(q_i^C\right)'[\kappa]}{1 + q_i^C[\kappa]}
$$

$$
- \frac{\left(x_i^T + x_i^C\right)\left(q_i^C\right)'[\kappa]}{q_i^C[\kappa]} + \frac{(n_i^C - x_i^C)x_i^T - x_i^T \sqrt{r_i(\kappa)}}{\kappa \sqrt{r_i(\kappa)}}
$$

$$
\text{and } r_i(\kappa) \quad = \quad -4(x_i^T + x_i^C)(x_i^T + x_i^C - n_i^T - n_i^C)\kappa
$$
$$
+ (x_i^T + x_i^C - n_i^C + (x_i^T + x_i^C - n_i^T)\kappa)^2
$$

is started with $\theta = 1$ and the result of the first iteration step is identical with the Mantel-Haenszel estimator, since

$$
\Gamma(1) = \frac{\sum_{i=1}^{n} \frac{x_i^T (n_i^C - x_i^C)}{\sqrt{r_i(1)}}}{\sum_{i=1}^{n} t_i(1)} = \frac{\sum_{i=1}^{n} \frac{x_i^T (n_i^C - x_i^C)}{n_i^C + n_i^T}}{\sum_{i=1}^{n} \frac{x_i^C (n_i^T - x_i^T)}{(n_i^C + n_i^T)}} = \kappa_{MH}
$$

as we have that

$$
r_i(1) = -4 \left(x_i^T + x_i^C\right) \left(x_i^T + x_i^C - n_i^T - n_i^C\right)
$$
$$
+ \left(x_i^T + x_i^C - n_i^C + \left(x_i^T + x_i^C - n_i^T\right)\right)^2
$$
$$
= \left(n_i^T + n_i^C\right)^2,
$$

and also

$$
t_i(1) = \frac{n_i^T \left(x_i^T + x_i^C\right) \left(x_i^C + x_i^T - n_i^T \left(n_i^C + n_i^T\right)\right)}{-2 \left(n_i^C + n_i^T\right) \left(x_i^C + x_i^T - n_i^C - n_i^T\right)}
$$
$$
+ \frac{n_i^T \left(x_i^T + x_i^C\right) \left(x_i^C + x_i^T - n_i^C - n_i^T\right)}{\left(n_i^C + n_i^T\right) \left(x_i^C + x_i^T - n_i^C - n_i^T\right)}
$$
$$
+ \frac{n_i^C \left(x_i^C + x_i^T\right) \left(x_i^C + x_i^T - n_i^T \left(n_i^C + n_i^T\right)\right)}{-2 \left(n_i^C - n_i^T\right) \left(x_i^C + x_i^T - n_i^C - n_i^T\right)}
$$
$$
+ \frac{\left(x_i^C + x_i^T\right) \left(x_i^C + x_i^T - n_i^T \left(n_i^C + n_i^T\right)\right)}{2(x_i^C + x_i^T - n_i^C - n_i^T)}
$$
$$
+ \frac{\left(n_i^C - x_i^C\right) x_i^T - x_i^T \left(n_i^C + n_i^T\right)}{\left(n_i^C + n_i^T\right)}
$$

$$= \frac{\left(x_i^C + x_i^T\right)\left(x_i^C + x_i^T - n_i^T\left(n_i^C + n_i^T\right)\right)}{-2(x_i^C + x_i^T - n_i^C - n_i^T)} + \frac{n_i^T\left(x_i^T + x_i^C\right)}{\left(n_i^C + n_i^T\right)}$$

$$+ \frac{\left(x_i^C + x_i^T\right)\left(x_i^C + x_i^T - n_i^T\left(n_i^C + n_i^T\right)\right)}{2(x_i^C + x_i^T - n_i^C - n_i^T)} + \frac{\left(n_i^C - x_i^C\right)x_i^T - x_i^T\left(n_i^C + n_i^T\right)}{\left(n_i^C + n_i^T\right)}$$

$$= \frac{n_i^T\left(x_i^T + x_i^C\right) + \left(n_i^C - x_i^C\right)x_i^T - x_i^T\left(n_i^C + n_i^T\right)}{\left(n_i^C + n_i^T\right)}$$

$$= \frac{n_i^T x_i^T + n_i^T x_i^C + n_i^C x_i^T - x_i^C x_i^T - x_i^T n_i^C - x_i^T n_i^T}{\left(n_i^C + n_i^T\right)}$$

$$= \frac{x_i^C\left(n_i^T - x_i^T\right)}{\left(n_i^C + n_i^T\right)}.$$

Bibliography

Agresti, A. and J. Hartzel (2000). Strategies for comparing treatments on binary response with multicenter data. *Statistics in Medicine 19*, 1115–1139.

Aitkin, M. (1998). Profile likelihood. In P. Armitage and T. Colton (Eds.), *Encyclopedia of Biostatistics*, Volume 5, pp. 3534–3536. Chichester, New York: Wiley.

Aitkin, M. (1999a). A general maximum likelihood analysis of variance components in generalized linear models. *Biometrics 55*, 117–128.

Aitkin, M. (1999b). Meta-analysis by random effect modelling in generalized linear models. *Statistics in Medicine 18*, 2343–2351.

Antiplatelet Trialists' Collaboration (1988). Secondary prevention of vascular disease by prolonged antiplatelet treatment. *British Medical Journal 296*, 320–332.

Arends, L. R., A. W. Hoes, J. Lubsen, D. E. Grobbee, and T. Stijnen (2000). Baseline risk as predictor of treatment benefit: Three clinical meta-re-analyses. *Statistics in Medicine 19*, 3497–3518.

Berkey, C., D. Hoaglin, F. Mosteller, and G. Colditz (1995). A random-effects regression model for meta-analysis. *Statistics in Medicine 14*, 395–411.

Berry, S. M. (1998). Understanding and testing for heterogeneity across 2x2 tables: Application to meta-analysis. *Statistics in Medicine 17*, 2353–2369.

Biggerstaff, B. J. and R. L. Tweedie (1997). Incorporating variability in estimates of heterogeneity in the random effects model in meta-analysis. *Statistics in Medicine 16*, 753–768.

Bird, S. M. (2003). European union's rapid TSE testing in adult cattle and sheep: Implementation and results in 2001 and 2002. *Statistical Methods in Medical Research 12*, 261–278.

Böhning, D. (1992). Multinomial logistic regression algorithm. *Annals of the Institute of Statistical Mathematics 44*, 197–200.

Böhning, D. (2000). *Computer-Assisted Analysis of Mixtures and Applications. Meta-Analysis, Disease Mapping and Others*. Boca Raton: Chapman & Hall/CRC.

Böhning, D. (2003). The EM algorithm with gradient function update for discrete mixtures with known (fixed) number of components. *Statistics and Computing 13*, 257–265.

Böhning, D. and B. G. Lindsay (1988). Monotonicity of quadratic approximation algorithms. *Annals of the Institute of Statistical Mathematics 40*, 223–224.

Böhning, D., U. Malzahn, E. Dietz, P. Schlattmann, C. Viwatwongkasem, and A. Biggeri (2002). Some general points in estimating heterogeneity variance with the DerSimonian-Laird estimator. *Biostatistics 3*, 445–457.

Böhning, D., J. Sarol, S. Rattanasiri, and A. Biggeri (2002). Efficient non-iterative and nonparametric estimation of heterogeneity variance for the standardized mortality ratio. *Annals of Institute of Statistical Mathematics 54*, 827–839.

Böhning, D., J. Sarol, S. Rattanasiri, C. Viwatwongkasem, and A. Biggeri (2004). A comparison of non-iterative and iterative estimators of heterogeneity variance for the standardized mortality ratio. *Biostatistics 5*, 61–74.

Brensen, R. M. D., M. J. A. Tasche, and N. J. D. Nagelkerke (1999). Some notes on baseline risk and heterogeneity in meta-analysis. *Statistics in Medicine 18(2)*, 233–238.

Breslow, N. E. (1984). Elementary methods of cohort analysis. *International Journal of Epidemiology 13*, 112–115.

Brockwell, S. E. and I. R. Gordon (2001). A comparison of statistical methods for meta-analysis. *Statistics in Medicine 20*, 825–840.

Brown, H. K. and R. J. Prescott (1999). *Applied Mixed Models in Medicine*. Chichester, West Sussex, England: Wiley.

Bryant, J., B. Fisher, N. Gunduz, J. P. Costantino, and B. Emir (1998). S-phase fraction combined with other patient and tumor characteristics for the prognosis of node-negative, estrogen-receptor positive breast cancer. *Breast Cancer Research and Treatment 51*, 239–253.

Cochran, W. G. (1954). The combination of estimates from different experiments. *Biometrics 10*, 101–129.

Cochrane Library (2005). *The Cochrane Collaboration*. http://www3.interscience.wiley.com/aboutus/sharedfiles/cochrane _transition: The Cochrane Library.

Colditz, G., T. Brewer, C. Berkey, M. Wilson, H. Fineberg, and F. Mosteller (1994). Efficacy of BCG vaccine in the prevention of tuberculosis. meta-analysis of the published literature. *Journal of the American Medical Association 271*, 698–702.

Cooper, H. and L. V. Hedges (1994). *The Handbook Research Synthesis*. New York: Russell Sage Foundation.

Cooper, M. R., K. B. G. Dear, O. R. McIntyre, H. Ozer, J. Ellerton, G. Can-
nellos, B. Duggan, and C. Schiffer (1993). A randomized clinical trial comparing Melphalan/Prednisone with and without a-2b interferon in newly-
diagnosed patients with multiple myeloma: A cancer and leukemia group b
study. *Journal of Clinical Oncology 11*, 155–160.

Davey Smith, G., F. Song, and T. A. Sheldon (1993). Cholesterol lowering
and mortality: The importance of considering initial level of risk. *British
Medical Journal 306*, 1367–1373.

Del Rio Vilas, V. J., P. Hopp, T. Nunes, G. Ru, K. Sivam, and A. Ortiz-
Pelaez (2007). Explaining the heterogeneous scrapie surveillance figures
across europe: a meta-regression approach. *BMC Veterinary Research 3*,
13.

Del Rio Vilas, V. J., J. Ryan, H. G. Elliott, S. C. Tongue, and J. W. Wilesmith
(2005). Prevalence of scrapie in sheep: Results from fallen stock surveys in
Great Britain in 2002 and 2003. *Veterinary Record 157*, 744.

Demidenko, E. (2004). *Mixed Models. Theory and Applications*. Hoboken,
New Jersey: Wliey.

Dempster, A., N. Laird, and D. B. Rubin (1977). Maximum likelihood estimation from incomplete data via the EM algorithm (with discussion). *Journal
of the Royal Statistical Society B 39*, 1–38.

DerSimonian, R. and N. Laird (1986). Meta-analysis in clinical trials. *Controlled Clinical Trials 7*, 177–188.

DuMouchel, W. and S.-L. T. Normand (2000). Computer-modeling and graph-
ical strategies for meta-analysis. In D. K. Stangl and D. A. Berry (Eds.),
Meta-Analysis in Medicine and Health Policy, pp. 127–178. Basel: Marcel
Dekker.

Efron, B. (1993). *An Introduction to the Bootstrap*. London: Chapman &
Hall.

Egger, M. and G. D. Smith (1995). Risks and benefits of treating mild hypertension: A misleading meta-analysis? [comment]. *Journal of Hypertension 13(7)*, 813–815.

Engels, E. A., C. H. Schmid, N. Terrin, I. Olkin, and J. Lau (2000). Heterogeneity and statistical significance in meta-analysis: An empirical study of
125 meta-analyses. *Statistics in Medicine 19*, 1707–1728.

European Commission (2001). Commission regulation (EC) no 1248/2001 of
22 june 2001 amending annexes III, x and XI to regulation (EC) no 999/2001
of the European Parliament and of the council as regards epidemio-
surveillance and testing of transmissible spongiform encephalopathies. *Of-
ficial Journal of the European Communities 44*, 12–22.

European Commission (2004). Report on the monitoring and testing of
ruminants for the presence of transmissible spongiform encephalopathy
(TSE) in the EU in 2003, including the results of the survey of the

prion protein genotypes in sheep breeds. Technical report, Brussels, http://europa.eu.int/comm/foot/food/biosafty/bse/annual_report_tse2003 _en.pdf, accessed on 10.01.2007.

European Stroke Prevention Study Group (1987). The european stroke prevention study (ESPS) principal endpoints. *Lancet 2*, 1351–1354.

Foster, J. D., D. Parnham, A. Chong, W. Goldmann, and N. Hunter (2001). Clinical signs, histopathology and genetics of experimental transmission of BSE and natural scrapie to sheep and goats. *Veterinary Record 148*, 165–171.

Greenland, S. (1994). A critical look at some popular meta-analysis methods. *American Journal of Epidemiology 140*, 290–296.

Greenland, S. and J. M. Robins (1985). Estimation of common effect parameter from sparse follow up data. *Biometrics 41*, 55–68.

Greiner, M. (2000). *Serodiagnostische Tests*. Berlin, Heidelberg, New York: Springer.

Hall, S., R. J. Prescott, R. J. Hallam, S. Dixon, R. E. Harvey, and S. G. Ball (1991). A comparative study of carvedilol, slow release nidedipine and atenolol in the management of essential hypertension. *Journal of Pharmacology 18*, S36–S38.

Hardy, R. J. and S. G. Thompson (1996). A likelihood approach to meta-analysis with random effects. *Statistics in Medicine 15*, 619–629.

Hardy, R. J. and S. G. Thompson (1998). Detecting and describing heterogeneity in meta-analysis. *Statistics in Medicine 17*, 841–856.

Hartung, J., D. Argac, and K. Makambi (2003). Homogeneity tests in meta-analysis. In R. Schulze, H. Holling, and D. Böhning (Eds.), *Meta-Analysis: New Developments and Applications in Medical and Social Sciences*, pp. 3–20. Göttingen: Hogrefe & Huber.

Hartung, J. and G. Knapp (2003). An alternative test procedure for meta-analysis. In R. Schulze, H. Holling, and D. Böhning (Eds.), *Meta-Analysis: New Developments and Applications in Medical and Social Sciences*, pp. 53–68. Göttingen: Hogrefe & Huber.

Hedges, L. V. (1994a). Combining estimates across studies: Meta-analysis of research. In *Clinical Practice Guideline Development: Methodology Perspectives*, pp. 15–26. Washington D. C: Agency for Health Care Policy Research.

Hedges, L. V. (1994b). Fixed effects model. In H. Cooper and L. V. Hedges (Eds.), *The Handbook of Research Synthesis*, pp. 286–299. New York: Russell Sage Foundation.

Hedges, L. V. and J. L. Vevea (1998). Fixed- and random-effects models in meta-analysis. *Psychological Methods 3(4)*, 486–504.

Higgins, J. P. T. and S. G. Thompson (2002). Quantifying heterogeneity in a meta-analysis. *Statistics in Medicine 21*, 1539–1558.

Hine, L. K., N. Laird, P. Hewitt, and T. C. Chalmers (1989). Meta-analytic evidence against prophylactic use of lidocaine in myocardial infarction. *Archives of Internal Medicine 149*, 2694–2698.

Hoes, A. W., D. E. Grobbee, and J. Lubsen (1995). Does drug treatment improve survival? Reconciling the trials in mild-to-moderate hypertension. *Journal of Hypertension 13*, 805–811.

Horwitz, R. I., B. H. Singer, R. W. Makuch, and C. M. Viscoli (1996). Can treatment that is helpful on average be harmful to some patients? A study of the conflicting information needs of clinical inquiry and drug regulation. *Journal of Clinical Epidemiology 49*, 395–400.

Hunter, N. (2003). Scrapie and experimental BSE in sheep. *British Medical Bulletin 66*, 171–183.

Jackson, D. (2006). The power of standard test for the presence of heterogeneity in meta-analysis. *Statistics in Medicine 25*, 2688–2699.

Jones, D. R. (1992). Meta-analysis of observational epidemiological studies: A review. *Journal of the Royal Society of Medicine 85(3)*, 165–168.

Kiefer, J. and J. Wolfowitz (1956). Consistency of the maximum likelihood estimator in the presence of infinitely many incidental parameters. *Annals of Mathematical Statistics 27*, 886–906.

Knapp, G., B. J. Biggerstaff, and J. Hartung (2006). Assessing the amount of heterogeneity in random-effects meta-analysis. *Biometrical Journal 48*, 271–285.

Kuhnert, R. (2005). *Untersuchung von Verschiedenen Modellierungen der Heterogenität in Multizentrischen Studien*. Ph. D. thesis, Charité Medical School.

Lachin, J. M. (2000). *Biostatistical Models - The Assessment of Relative Risks*. New York: Wiley & Sons.

Laird, N. (1978). Nonparametric maximum likelihood estimation of a mixing distribution. *Journal of the American Statistical Association 73*, 805–811.

Lange, K. (2004). *Optimization*. New York, Berlin, London: Springer.

Le, C. T. (1992). *Fundamentals of Biostatistical Inference*. New York, Basel, Hong Kong: Marcel Dekker, Inc.

Lindsay, B. G. (1983). The geometry of mixture likelihoods, part i: A general theory. *Annals of Statistics 11*, 783–792.

Lindsay, B. G. (1995). *Mixture Models: Theory, Geometry, and Applications*. Institute of Statistical Mathematics, Hayward: NSF-CBMS Regional Conference Series in Probability and Statistics.

Lipsitz, S. R., K. B. G. Dear, N. M. Laird, and G. Molenberghs (1998). Tests for homogeneity of the risk difference when data are sparse. *Biometrics 54*, 148–160.

Malzahn, U., D. Böhning, and H. Holling (2000). Nonparametric estimation of heterogeneity variance for the standardised difference used in meta-analysis. *Biometrika 87*, 619–632.

Marshall, R. J. (1991). Mapping disease and mortality rates using empirical Bayes estimators. *Applied Statistics 40*, 283–294.

Martuzzi, M. and M. Hills (1995). Estimating the degree of heterogeneity between event rates unsing likelihood. *American Journal of Epidemiology 141*, 369–374.

McCullagh, P. and J. A. Nelder (1989). *Generalized Linear Models* (2nd ed.). London: Chapman & Hall.

McLachlan, G. and T. Krishnan (1997). *The EM Algorithm and Extensions*. New York: Wiley.

McLachlan, G. and D. Peel (2000). *Finite Mixture Models*. New York: Wiley.

Murphy, S. A. and A. W. Van der Vaart (2000). On profile likelihood. *Journal of the American Statistical Association 95*, 449–485.

Nass, C. A. G. (1959). The chi-square test for small expectations in contingency tables, with special reference to accidents and absenteeism. *Biometrika 46*, 365–385.

Nelder, J. A. and R. W. M. Wedderburn (1972). Generalized linear models. *Journal of the Royal Statistical Society, Series A 135*, 370–384.

Neyman, J. and E. L. Scott (1948). Consistent estimates based on partially consistent observations. *Econometrica 16*, 1–32.

Normand, S. L. T. (1999). Tutorial in biostatistics: Meta-analysis: Formulating, evaluating, combing, and reporting. *Statistics in Medicine 18*, 321–359.

Pawitan, Y. (2001). *In All Likelihood. Statistical Modelling and Inference Using Likelihood*. Oxford: Clarendon Press.

Petitti, D. B. (1994). *Meta-Analysis, Decision Analysis and Cost - Effectiveness Analysis. Methods for Quantitative Synthesis in Medicine*. Oxford: Oxford University Press.

Pocock, J. S. (1997). *Clinical Trials: A Practical Approach*. Chichester, New York, Brisbane: John Wiley & Sons.

Potthoff, R. F. and M. Whittinghill (1966). Testing for homogeneity i: The binomial and multinomial distributions. *Biometrika 53*, 167–182.

Raudenbush, S. W. (1994). Random effects models. In H. Cooper and L. V. Hedges (Eds.), *The Handbook of Research Synthesis*, pp. 302–321. New York: Russell Sage Foundation.

Scientific Steering Committee (2001). *Opinion on Requirements for Statistically Authoritative BSE/TSE Surveys. Adopted by the Scientific Steering Committee at its Meeting of 29-30 November 2001*. European Commission, Health & Consumer Protection Directorate-General. http://ec.europa.eu/food/fs/sc/ssc/out238_en.pdf, accessed on 10.01.2007.

Selective Decontamination of the Digestive Tract Trialists' Collaborative Group (1993). Meta-analysis of randomised controlled trials of selective decontamination of the digestive tract. *British Medical Journal 307*, 525–532.

Sharp, S. J. and S. G. Thompson (2000). Analysing the relationship between treatment effect and underlying risk in meta-analysis: Comparison and development of approaches. *Statistics in Medicine 19*, 3251–3274.

Sharp, S. J., S. G. Thompson, and D. G. Altman (1996). The relation between treatment benefit and underlying risk in meta-analysis. *British Medical Journal 313(7059)*, 735–738.

Sidik, K. and J. N. Jonkman (2005). Simple heterogeneity variance estimation for meta-analysis. *Journal of Applied Statistical Science 54*, 367–384.

StataCorp. (2005). Stata statistical software: Release. College Station, TX: StataCorp LP.

Stram, D. O. (1996). Meta-analysis of published data using a linear mixed-effects model. *Biometrics 52*, 536–544.

Sutton, A. J., K. R. Abrams, D. Jones, T. A. Sheldon, and F. Song (2000). *Methods for Meta-Analysis in Medical Research*. New York: John Wiley & Sons, LTD.

Thompson, S. G. (1993). Controversies in meta-analysis: The case of the trials of serum cholesterol reduction. *Statistical Methods in Medical Research 20*, 173–192.

Thompson, S. G. (1994). Why sources of heterogeneity in meta-analysis should be investigated. *British Medical Journal 309(6965)*, 1351–1355.

Thompson, S. G. and S. J. Sharp (1999). Explaining heterogeneity in meta-analysis: A comparison of methods. *Statistics in Medicine 18*, 2693–2708.

Turner, R., R. Omar, M. Yang, H. Goldstein, and S. Thompson (2000). A multilevel model framework for meta-analysis of clinical trials with binary outcome. *Statistics in Medicine 19*, 3417–3432.

van Houwelingen, H. C., L. R. Arends, and T. Stijnen (2002). Tutorial in biostatistics. Advanced methods in meta-analysis: Multivariate approach and meta-regression. *Statistics in Medicine 21*, 589–624.

van Houwelingen, H. C. and S. Senn (1999). Investigating underlying risk as a source of heterogeneity in meta-analysis (letter). *Statistics in Medicine 18*, 107–113.

van Houwelingen, H. C., K. Zwinderman, and T. Stijnen (1993). A bivariate approach to meta-analysis. *Statistics in Medicine 12*, 2272–2284.

Whitehead, A. (2002). *Meta-Analysis of Controlled Clinical Trials*. Chichester: Wiley.

Whitehead, A. and J. Whitehead (1991). A general parametric approach to the meta-analysis of randomised clinical trials. *Statistics in Medicine 10*, 1665–1677.

Wilesmith, J. W., J. Ryan, V. J. Del Rio Vilas, and S. Gubbins (2004). Summary of the results of scrapie surveillance in sheep in Great Britain, April-December 2003. *Weybridge: Veterinary Laboratories Agency.* http://www.defra.gov.uk/animalh/bse/othertses/scrapie/scrapiesurvey.pdf, accessed on 10.01.2007.

Wilesmith, J. W., G. A. H. Wells, M. P. Cranwell, and J. B. M. Ryan (1988). Bovine spongiform encephalopathy: Epidemiological studies. *Veterinary Record 123*, 638–644.

Will, R. G., J. W. Ironside, M. Zeidler, S. N. Cousens, K. Estibeiro, A. Alperovitch, S. Poser, M. Pocchiari, A. Hofman, and P. G. Smith (1996). A new variant of Creutzfeldt-Jakob disease in the UK. *Lancet 347*, 921–925.

Woodward, M. (1999). *Epidemiology: Study Design and Data Analysis.* London, New York, Washington, D.C.: Chapman & Hall/CRC.

Yusuf, S., R. Peto, J. Lewis, R. Collins, and P. Sleight (1985). Beta blockade during and after myocardial infarction: An overview of the randomized trials. *Progress in Cardiovascular Diseases 27*, 335–371.

Author index

Subject index

T - #0439 - 071024 - C14 - 234/156/9 - PB - 9780367387570 - Gloss Lamination